CAMBRIDGE COUNTY GEOGRAPHIES

General Editor: F. H. H. GUILLEMARD, M.A., M.D.

EAST LONDON

Cambridge County Geographies

EAST LONDON

by

G. F. BOSWORTH, F.R.G.S.

With Maps, Diagrams and Illustrations

Cambridge:

at the University Press

1911

CAMBRIDGE UNIVERSITY PRESS
Cambridge, New York, Melbourne, Madrid, Cape Town,
Singapore, São Paulo, Delhi, Mexico City

Cambridge University Press
The Edinburgh Building, Cambridge CB2 8RU, UK

Published in the United States of America by Cambridge University Press, New York

www.cambridge.org
Information on this title: www.cambridge.org/9781107667501

First published 1911
First paperback edition 2013

A catalogue record for this publication is available from the British Library

ISBN 978-1-107-66750-1 Paperback

CONTENTS

ILLUSTRATIONS

The illustrations on pp. 6, 35, 42, 72, 100, 101, 104, 105, 109, 111, 121, 138, 139, 141, 143, 146, 149, 151, 155, 168, 171, 175, 180, 190, 198, 204, 206, 207, 215, 216 and 244 are from photographs by Messrs Frith & Co.; those on pp. 28, 33, 50, 57, 88, 97, 157 and 214 from photographs by Messrs J. Valentine & Sons, Ltd.; those on pp. 78, 84, 95, 136 and 193 from photographs by Mr Bridgen; those on pp. 61 and 158 from photographs by the Sport and General Illustrations Co.; that on p. 87 is reproduced by kind permission of *The Times*; that on p. 74 is from a photo kindly supplied by Sir W. Christie; those on pp. 22 and 241 are from photos by Mr A. Wire; those on pp. 220, 226 and 234 are from photographs by Mr Emery Walker; those on pp. 63, 116, 188, 229 and 230 are reproduced from *Literary London* by the courtesy of Mr T. Werner Laurie; the design on p. 200 is reproduced by kind permission of the *Building News*.

1. County and Shire. The County of London. The word *London*. Its Origin and Meaning.

The main divisions of our country are known as counties, and, in some instances, as shires. When the word shire is used, it is added to the county name. For instance, we speak of the county of Kent, or of the county of Bedford ; but while the word shire is not added to the name of Kent, it may be to that of Bedford. Thus we write the county of Bedford, or Bedfordshire, but not the county of Bedfordshire. Such an expression would be superfluous, for the word shire is now practically equivalent to the later word county.

Although, however, we now call all the divisions of England and Wales counties, that title is not historically accurate. Some counties, such as Kent, Essex, and Sussex, are really survivals of various old English kingdoms, and for more than a thousand years there has been but little alteration either in their boundaries or their names.

The divisions now known as Bedfordshire, Hertfordshire, and Wiltshire are so called because they were *shares*

or portions *shorn off* from larger kingdoms.   Thus Bedfordshire and Hertfordshire were shares or portions of a very
large kingdom known as Mercia, while Wiltshire was a
share or portion of Wessex.   It is not necessary to enlarge
further on this distinction, but it is well to have a correct
idea of the origin of our counties.   For many years it was
wrongly stated that Alfred divided England into counties.
The statement is incorrect, for we know that some of the

Entrance to London at Mile End in the Eighteenth Century

counties were in existence before his time, while others
were formed after his death.

It may be stated here that the object of thus dividing
our country into counties was partly military and partly
financial.   Every shire had to provide a certain number
of armed men to fight the king's battles, and also to pay
a certain proportion of the king's income.   In each case a
"shire-reeve," or sheriff as we now call him, was appointed

by the king to see that the shire did its duty in both respects. After the Norman Conquest, the government of each shire was handed over to a count, and from that time these divisions have been called counties.

In England the divisions or ancient counties numbered forty until the year 1888. Then it was decided to form the Administrative County of London, under the provisions of the Local Government Act of that year. It is to be noted that although London is the latest of the forty-one counties, it is not known as an "ancient" county, for it was constituted an administrative area from parts of the ancient counties of Middlesex, Surrey, and Kent. Thus it comes about that London, the capital of the British Empire, the greatest city in the world, and once the capital of the county of Middlesex, is now an Administrative County with an enormous population.

There is another London, which is often called "Greater London"; but with that we do not propose to deal, as that enlarged area takes in many parishes and districts that are outside the boundaries of the administrative county and extend into Hertfordshire and Essex.

Now with regard to the name *London*, there is great diversity of opinion as to its origin and meaning. We shall not, however, be wrong if, in giving some of the opinions on this subject, we state that the earliest historic monument of London is its name. The word *Londinium* first appears in Tacitus under the year A.D. 61 as that of an *oppidum* not dignified with the name of a colony, but celebrated for the gathering of dealers and commodities.

It follows from this early notice that *Londinium* must have been founded long before A.D. 61. Historians have come to the conclusion that the Roman *oppidum* was built on the site of an earlier Celtic village, and that the name *Londinium* is probably the Latinised form of *Llyn-Din*, i.e. the lake-fort.

The Arms, Crest, and Supporters of the City of London

Some writers have endeavoured to explain the name from other Welsh roots, but nothing is so uncertain as the origin of some place-names. Geoffrey of Monmouth thinks that London was called *Caer-Lud* after a King Lud of Celtic history, and even some recent writers have come round to this view and say that London means

Lud's-town. This last derivation may be mere conjecture, although it is in harmony with tradition.

It may be mentioned that Geoffrey of Monmouth
wrote early in the twelfth century, and gives a legend
of the founding of London. This describes how Brutus
came over from Troy and formed the plan of building a
city. When he came to the Thames he found a site on
its banks most suitable for this purpose. There he built
a city calling it *Troia Nova*, i.e. New Troy, which was
afterwards corrupted into *Trinovantum*. As time passed
on, King Lud built walls and towers round the city; and,
when he died, his body was buried by the gate which is
called in the Celtic speech " Porthlud," but in the Saxon
" Ludesgata."

Here, then, we have the legend of the origin of
London in pre-Roman days, and it may be founded on
some genuine folk-stories of Celtic origin. At any rate, it
explains the fact that the Roman attempt to change the
name to *Augusta* completely failed, for the early name
Llyn-din, or Caer-Lud, held its own in the affections
of the Britons. Whatever conclusion we reach with
regard to the origin of the name London, we feel sure
that it was a village of some importance before the
Roman occupation, as prehistoric and early relics are
often found on the site.

Thus it comes about that London has an almost
unbroken record extending over 2000 years, and whether
as Llyn-din, or Augusta, or Londinium, or London,
occupies a commanding place in our country's history.

2. General Characteristics.   Position and Natural Conditions.   Why London is our capital.

There may be doubts as to the origin of London and the exact meaning of its name, but there can be no doubt as to its two thousand years of unbroken history and that

The Embankment looking Citywards from Charing Cross

it exerts a great fascination over the imagination of Englishmen. It has been well remarked that "London has a charm all her own ; it is that of a history as romantic and as interesting to Englishmen as that of Ancient Rome was to the Romans. As Ancient Rome once was, so is London now the centre of civilisation."

In this chapter we shall do well if we first glance at some of the general characteristics of London, and then pass on to consider its position, and why it came to be chosen as our capital. There are people who would argue that London is a most unsuitable site for a capital, but we have to remember that it has stood the great practical test of centuries and has won its way to the foremost place against the competition of other cities that were officially favoured. Thus York was the chief Roman centre of administration, and Winchester was the chief city of Wessex and became the capital when the Kings of Wessex were supreme over all England.

It is sometimes easy to give the characteristics of a city or of some place of historic interest. But in dealing with London we have to think of at least two cities, round which have grown numerous towns that would each be considered large in the provinces. The vastness of London is so overwhelming, and its variety so amazing, that we are not surprised to find how differently it is characterised by poets and historians.

Wordsworth was charmed with the sight of London from Westminster Bridge, and in one of his sonnets exclaims :—

> " Earth has not anything to show more fair :
> Dull would he be of soul who could pass by
> A sight so touching in its majesty " :

Byron looked upon it as " a mighty mass of brick, and smoke, and shipping." A French writer calls it " a province in brick " ; and one of our own literary men characterises it as " a squalid village." Heine, the great

German writer, gives his idea of London as "a forest of houses, between which ebbs and flows a stream of human faces, with all their varied passions—an awful rush of love, hunger, and hate."

There is some truth in each of these various attempts to give an idea of London, but of course they are all short of conveying the correct impression. Probably no one man is capable of giving a true picture of London, for there are so many aspects of the modern city. Its immense population and the strange variety of races is sure to have its effect on one class of observers. Another class of people will be struck with the contrasts between the princely palaces of the rich and the filthy hovels of the poor, or between the magnificent squares and the squalid slums. In no other city in the world is there such a striking contrast between historic buildings which date from the Conquest and the modern structures of stone and marble which have supplanted the wooden houses of the Stuart period.

Such, then, are a few of the most remarkable characteristics of London as it is to-day. It is not possible to deal further with this subject in the present book, so we will proceed to consider the position of London and what effect the choice of the site of the City by the early founders had on its subsequent prosperity.

It will be well to look at an early map of the capital showing the marshes on either side of the Thames. We shall then get some idea of what the Thames was like in British days. Then the river must have looked like a broad lake, with here and there a small island rising out of the water. When the tide was high, the river was converted into an arm of the sea, while at low water it was a vast marsh through which the stream wound its way in irregular fashion. It has been estimated that at

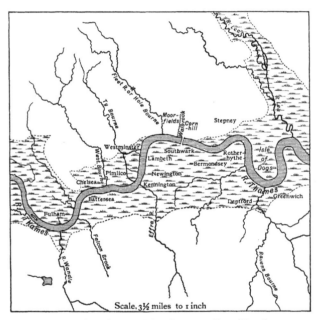

Ancient London and its Surrounding Marshes

least half of modern London is built on this marsh, which extended from Fulham on the west to Greenwich on the east.

In those far-off days, the marsh was the resort of wild duck, wild geese, herons, and other water birds, which flew

over it in myriads.   Altogether we can picture the site
of London two thousand years ago as a dreary and desolate
place, and one of the first questions that arises from this
knowledge is, How came London to be founded on a marsh?

There are many reasons why London was founded
on the present site, and if we consider a few of them it
will help us to understand its growth and development.
Of course we are referring to the site of London as it was
in the time of the British founders, or at the period of
the Roman Conquest.   The evidence goes to show that
the earliest centre of the City was on the east side of the
Walbrook at the head of London Bridge.   Now taking
that district as the nucleus of the early city, we find that
London was built on the first spot going up the river
where any tract of dry land touches the stream.   We
also find that it is a tract of good gravel soil, well supplied
with water and not liable to flooding.   These were most
important considerations in selecting the site of a city in
those early days, just as they are at the present time.

It will be seen that this spot of good land was chosen
on the river Thames, which was thus at once a means
of defence, and a highway which could be traversed both
up and down by means of the British boats.   The site
was not very near the sea, and that fact was also an
advantage, for the small boats of the Britons could not
venture on the waves of the Lower Thames.   There is
no doubt that the settlement was founded on a site about
60 miles from the coast, because it was not open to attack
from the enemies who came over the sea.   It is here
worth mentioning that *London* and *Thames* are both Celtic

words, and are the only names remaining in this area to remind us of the British occupation.

There is one other reason we may consider in this connection. London was placed on a tidal river, and thus it carried boats laden with merchandise or passengers far up the river to the west, and far down the river to the east. We may be sure that the Britons made use of the tide, and the Romans, who had been accustomed to the nearly tideless waters of the Mediterranean, soon learnt the value of the ebbing and flowing of the Thames.

Thus we may conclude that the earliest site of London was on land about fifty feet above the level of the tide, and the position was admirably adapted for defence, for it was almost impregnable. Green, in *The Making of England*, remarks that London, "sheltered to east and south by the lagoons of the Lea and the Thames, guarded to the westward by the deep cleft of the Fleet, saw stretching along its northern border the broad fen whose name has survived in our modern Moorgate....The 'dun' was in fact the centre of a vast wilderness. Beyond the marshes to the east lay the forest tract of southern Essex. Across the lagoon to the south rose the woodlands of Sydenham and Forest Hill, themselves but advance guards of the fastnesses of the Weald. To the north the heights of Highgate and Hampstead were crowned with forest masses, through which the boar and the wild-ox wandered without fear of man to the days of the Plantagenets. Even the open country to the west was but a waste. It seems to have formed the border-land between two British tribes who dwelt in Hertford and in Essex—its barren clays were given over to solitude by the usages of primeval war."

Besides the geographical reasons accounting for the greatness of London, there are also historical and political reasons for its prosperity and development. Bristol and Liverpool on the west, and Plymouth and Southampton on the south, are equally well placed, and have enjoyed exceptional facilities for the cultivation of foreign trade. But, while these and other towns have been fettered by the action of their feudal lords, London has had no over-lord but the king. The City has always had rule over its own district, and was not controlled by any outside power. Thus it comes about that London has distanced all rivals, such as York and Winchester, and now stands without a peer, the capital of the British Empire and the greatest city of the world.

3. Size. Boundaries. Development. History of Growth. London of the Romans, of the Saxons, of the Normans. Medieval London. Stuart London.

As we have already seen in a former chapter, England was formerly divided into 40 geographical counties, but in 1888 it was decided to form the Administrative County of London. The number of geographical counties is now 41; but England is also divided by the Local Government Act of 1888 into 50 Administrative Counties. Some

Plan of the City of London showing the Wards

of the larger counties were then divided into two or more portions, and thus the old idea of 40 counties has become obsolete, and we now speak not only of Sussex and Suffolk, but also of East Sussex and West Sussex, of East Suffolk and West Suffolk. It is well to make this point quite clear, so that we may understand London's position as a county.

Of the 41 geographical counties in England, London is the most recently formed, it is the most important, and it is the smallest in point of size. A reference to the diagrams at the end of the book will illustrate its area compared with that of England and Wales. London contains 74,839 acres or 116·9 square miles, and is thus about $\frac{1}{498}$ of England and Wales. The heart of the county is called the City of London, and is about one square mile in area.

A glance at the map of the County of London will show that it is an irregularly-shaped area that has grown from a British village of less than half a square mile to its present size. It is divided into two unequal areas by the many windings of the River Thames : the northern portion is entirely formed from Middlesex, while the southern portion has been taken from both Surrey and Kent. The northern portion contains about two-fifths of the entire area, but in many respects it is the more important of the two divisions.

The length of the county measured from Hammer-smith on the west to Plumstead on the east is about 17 miles, while the breadth from Holloway in the north to Streatham in the south is about 11 miles. It will be noticed that there is a small portion of the county on the Essex side of the Thames. This is known as North Woolwich, and before 1888 was part of Kent although actually in the county of Essex.

Except on the east side, where it is bounded for some miles by the River Lea, the boundaries are not physical. Middlesex forms the boundary on the north and partly on the west, while Surrey bounds it partly on the west and south, and Kent partly on the south and east.

Before we go further, it must here be stated that the present volume on the eastern portion of London includes all the district east of the boroughs of St Pancras, Holborn, Westminster, and Lambeth, and has an area of 41,832 acres. This eastern portion of London comprises 16 out of the 29 boroughs into which the county is divided. Nine boroughs in the eastern portion are on the north of the Thames, and the remaining seven lie south of that river. The southern portion is much larger than the northern portion, although it is not so important, for we must always remember that, for many centuries, London as a city was only built on the north bank of the Thames.

The line of division that is chosen for this volume is purely arbitrary, and is merely for purposes of convenience. In the eastern portion we get the City of London with its surrounding boroughs, and in the western portion we have the City of Westminster and its neighbouring boroughs. Lewisham in the western portion, and Woolwich in the eastern portion, are the largest boroughs; and Holborn in the western, and Finsbury in the eastern, are the smallest.

A View of LONDON as it appeared
before the dreadful Fire An. 1666

References

1 St Pauls
2 St Faiths
3 Temple
4 St Brides
5 St Andrews
6 Bugnards Cast.

7 St Sepulchres
8 Pauls Church
9 Guild hall
10 St Michaels
11 St Laurence Poultry
12 Old Swan

13 London Bridge
14 St Magnus Ingl.
15 Billingsgate
16 Custom House
17 Tower
18 Dr Wharf
19 St Olaves

20 St Mary overs
21 Winchester house
22 The Globe
23 The Bear Garden
24 Hampstead
25 Highgate
26 Iselington

Having given these facts and figures relating to the size of the present County of London, we may briefly trace its growth and development from the earliest times. It would be quite impossible within the limits of this book to go into details; but we can give a few ideas as to its size and condition at three or four turning-points in its history.

In British times we must fall back on conjecture, but we have also the aid of geography and geology. The facts that prove the condition of the earliest London are the waste, marshy ground, with little hills rising from the plains, and the dense forest to the north. The position of the town on the Thames proves the wisdom of those who chose the site, although the frequent overflowing of the river must have hindered its progress.

Under the Romans, the city became the chief residence of merchants and the great mart of trade. The Romans probably built a fort where the Tower now stands, and afterwards the walls surrounding the town were erected. Then Londinium took its proper place among the Roman cities of Britain, for it was on the high road to York and the starting-point of most of the Roman roads in Britain. The two chief events in the history of Roman London are the building of the bridge and the building of the wall. The exact date of the building of the wall cannot be given, but we know that in 350 A.D. it did not exist, while in 368 A.D. the town with its villas, its gardens, and its township, was enclosed. A reference to the map on p. 161 will show the circuit of the wall, with its gates and

fort.   London within the wall occupied an area of about
380 acres, and was about $3\frac{1}{4}$ miles in circumference.   This
Roman wall round London was of the utmost importance
in the history of the city, and even to this day it forms in
part the City boundary.

When the Roman legions left Britain, London had a
very mixed population of traders.   The inhabitants were
defenceless and at the mercy of the invader.   The Saxons
conquered the eastern portion of England, and named it
Essex.   London became the capital of the East-Saxon
kingdom.   Saxon London was a wooden city, surrounded
by walls, which probably marked the same enclosure as
the Roman city.   In the seventh century the city settled
down into a prosperous place, and was peopled by merchants
of many nations.   It was a free trading town, and it was
also the great mart of slaves.   In the eighth and ninth
centuries it was frequently harried and desolated by the
Danes, but the great turning-point in its history was in
886 A.D., when King Alfred restored it and introduced a
garrison of men for its defence.   From this year to the
present time, London has been in the front rank of our
cities, and at the Norman Conquest it became, without
a rival, the capital of England.   The further growth
and development of the city were now very marked, and
William I granted a charter to William the Bishop, and
Gosfrith the Portreeve, who is supposed to be Geoffrey
de Mandeville.

If we want to get further particulars of the growth of
London, we must refer to our literature of the fourteenth
and subsequent centuries.   London places are frequently

mentioned in *Piers Plowman*; while Occleve, Gower, and Chaucer are invaluable to the student of early London life. *London Lickpenny*, attributed to Lydgate, is a valuable record of London life at the end of the fourteenth century. In it are related the adventures of a poor Kentish man who went to London in search of justice, but could not find it for lack of money. Chaucer gives us many pictures of the London of his day, and the portraits of the pilgrims in the Prologue to *The Canterbury Tales* show us the men and women who were to be seen daily in the streets of London.

When we come down to the Stuart period, we find that London had about 150,000 people in the reign of James I, and in the reign of Charles II we are told that "the trade and very City of London removes westward, and the walled City is but one-fifth of the whole pile." Lord Macaulay made a special study of the state of London in 1685, and the following extract from his *History of England* gives a very picturesque account of the condition of the City more than two hundred years ago. He writes thus :—"Whoever examines the maps of London which were published towards the close of the reign of Charles the Second will see that only the nucleus of the present capital then existed. The town did not, as now, fade by imperceptible degrees into the country. No long avenues of villas, embowered in lilacs and laburnums, extended from the great centre of wealth and civilisation almost to the boundaries of Middlesex, and far into the heart of Kent and Surrey. In the east no part of the immense line of warehouses and artificial lakes which now stretches from the Tower to Blackheath had been projected. On the west, scarcely one of those stately piles of building which are inhabited by the noble and wealthy was in existence; and Chelsea...was a quiet country village with about a thousand inhabitants. On the north cattle fed, and sportsmen wandered with dogs and guns over the site of the borough of Marylebone, and over far the greater part of the space now covered by... Finsbury and the Tower Hamlets. Islington was almost a solitude; and poets loved to contrast its silence and repose with the din and turmoil of the monster London. On the south the capital is now connected with its suburb by several bridges, not inferior in magnificence and solidity to the noblest works of the Caesars. In 1685, a single line of irregular arches, overhung by piles of mean and crazy houses, and garnished, after a fashion worthy of the naked barbarians of Dahomey, with scores of mouldering heads, impeded the navigation of the river."

Lord Macaulay wrote this interesting sketch of Stuart London more than 60 years ago, when the population of the metropolis was under two millions. Since Macaulay's time London has increased enormously both in area and population, and the contrast between the early Victorian London and that of to-day is almost as striking as that drawn by the great Whig historian. Although a term has been put to its extent by the Act of 1888, its population has increased, and, as we shall see in subsequent chapters, its development in trade and commerce is also progressive.

4. London Parks, Commons, and Open Spaces in the North=east and South=east.

If we look at any map of London showing the parks, commons, and open spaces within its boundaries we shall at once realise that Londoners are very fortunate in being so well provided with municipal "lungs." The first idea of many people who do not know London is that the metropolis is nothing more than a wilderness of brick and mortar. This, we shall find, is far from being true ; and probably no other capital in the world has such extensive breathing spaces for its people as ours. The finest and largest parks are, as we might expect, in the western portion of the county ; but we must remember that the people in the north-east have Epping Forest, which, although in Essex, is yet maintained by the City Corporation and is known as London's playground.

We may now try to get a good idea of the extent of London's parks and open spaces, then we will consider some of their characteristics, and finally we will pass in review those that are situated in the eastern portion of the County of London.

The parks, commons, and open spaces within the County of London have an extent of 6,588 acres, or 8·8 per cent. of the entire area. They are owned and maintained by the Government, the City Corporation, the London County Council, the various Borough Councils, the Conservators of Putney and Wimbledon

Commons, the Metropolitan Public Gardens Association, and various other public bodies and persons. The London County Council and the City Corporation also own and maintain forests, parks, and open spaces outside the County, and in some instances we shall specially refer to these. The Government own and maintain Hyde Park,

View in Epping Forest

St James's Park, the Green Park, Kensington Gardens, Regent's Park, Greenwich Park, Woolwich Common, and other smaller spaces. The City Corporation own and maintain Highgate Wood, Queen's Park, Kilburn, within the County ; and Epping Forest, Burnham Beeches, and West Ham Park outside the County. The London County Council are responsible for Battersea Park,

Bostall Heath and Woods, Brockwell Park, Clapham Common, Hackney Marsh, Hampstead Heath, Victoria Park, Tooting Common, Wandsworth Common, Streatham Common, Wormwood Scrubs, and many other open tracts. They recently came into possession of Hainault Forest, a most beautiful piece of woodland in Essex ; and not a year passes without one or more parks and open spaces either being given to or bought by the London County Council. The various Borough Councils maintain such open spaces as disused burial grounds, recreation grounds, gardens in squares, and small commons.

We can realise what a boon all these parks and open spaces must be to London when we remember that many Londoners can never get far away from their place of work or home all the year round. To thousands of men, women, and children the parks and open spaces in the great city afford their only place of recreation and give them some idea of what the country is like. It has been a great advantage to London to have these open spaces for public resort, for there is no doubt that through them the love of Londoners for flowers and birds has been developed. Englishmen have justifiably gained a reputation as landscape gardeners, and in our London parks we may see some good examples of their art. It has been the aim of those who laid out the parks to make them as natural as possible. A walk in Regent's Park or Kensington Gardens will at once show what beautiful tracts of woodland they are, and what care has been displayed in preserving their natural characteristics. In most of the parks, certain portions have been laid

out as flower gardens, and the varied colours of the
tastefully-arranged beds form charming pictures for the
wearied citizen. Besides the fine trees and beautiful
flowers, the Parks also have the great attraction of bird
life, but of this we shall read in another chapter.

In some of the London parks, perhaps, there has been
a tendency to make too many straight rows and formal
walks, but this cannot be said of the commons, or of such
a tract as Hampstead Heath. The commons have a
distinct charm in their natural beauty and in their freedom,
as opposed to the artificial character and restrictions of
some of the parks. These commons are also part of the
history of the county, and take us back to the time when
the land was tilled in common. Not many years ago,
there was a desire to build over these commons; but of
late a better spirit is abroad, and now every effort is made
for the preservation of open spaces in and around London.

In this our eastern section of the County of London
we find that, north of the Thames, the following are
the largest open spaces—Clissold Park, Hackney Downs,
Hackney Marsh, and Victoria Park; and south of the
Thames, Greenwich Park, Woolwich Common, Black-
heath, Bostall Heath and Woods, Eltham Common,
Plumstead Common, Dulwich Park, and Southwark
Park.

Clissold Park, better known as Stoke Newington
Park, has the great attraction of the New River, which
winds round and through it. Bird and animal life is
specially provided for: there are some fine cedars, while
one of the thorns is said to be the oldest in England.

Hackney Downs is one of the lammas-lands of Hackney and has on its borders the handsome school of the Grocers' Company. Hackney Marsh, a great open space of about 340 acres on the borders of the river Lea, has a fine expanse of level surface, and affords excellent facilities for cricket and football. It has been declared to be the finest playground in the world.

Victoria Park is the chief park of the East End, and plays a most important part in brightening the lives of the teeming thousands who dwell in the squalid streets of this great district. The area of the park is so large that it is possible to provide for nearly every form of out-door amusement and recreation. There are special facilities for bathing and swimming, and numerous pitches for cricket and other games. A feature which is very popular with children is the introduction of animal and bird life into the park. An aviary contains a varied selection of English birds, while there are goats in a rockery by themselves, and an enclosure for deer. Part of the area of the Park, which has an extent of 217 acres, is laid out with walks, lakes, and gardens. On the lakes are swans and pleasure-boats, and in the middle of the park is a memorial fountain in the shape of a Gothic temple.

Greenwich Park is 185 acres in extent, and was laid out by a French landscape-gardener in the reign of Charles II. There are avenues of splendid Spanish chestnut trees, steep hills, and green valleys. On one of its hills stands Greenwich Observatory, and from the terrace in front of this building and from other elevated portions of the park, extensive and varied views over the river and over Epping and Hainault Forests may be obtained. This park is a favourite resort of Londoners, who come to admire the deer and the flowering hawthorns in May-time.

Blackheath is one of the oldest and one of the largest of London's open spaces, with an area of 267 acres. It forms an elevated plateau, fairly level except for the extensive gravel pits, which nature has transformed into grassy dells. The otherwise bare appearance of the heath is somewhat relieved by fine clumps of trees. Blackheath, which is crossed by Watling Street, the Roman road from Dover to London, has been the scene of many great events in our history. It was in olden times a noted place for highway robberies, and fairs were formerly held on the Heath. It is now chiefly associated with golf, established here by James I as far back as 1608, and football, and with the enjoyment of Londoners on Bank-holidays.

Woolwich Common has an area of 159 acres, and is the property of the Government. It is used for exercising the troops and for reviews; but there is an open road across it, and the public have free access to it except when any part is required for military purposes. Only separated from it by the road is Eltham Common—a misleading name, for it is a long distance from Eltham and close to Woolwich. It has an area of about 40 acres and belongs to the War Office.

Further to the east is Plumstead Common, a fine open space of 103 acres. It is a wide-stretching, elevated plateau, broken here and there by slight depressions. Much of the common is flat and bare, but some parts are very beautiful, especially the deep valley at the east end known as the Slade.

Beyond Plumstead Common are Bostall Heath and Woods (see p. 57). These open spaces of 133 acres are most attractive and, with the exception of Epping Forest, are the most sylvan in character of all London's parks and commons. They are almost unknown to the great mass of Londoners, although they are a favourite resort of those who live near them. They are of considerable height and are crowned by stretches of pines and larches, while at every opening there are thick clumps of gorse and bracken, and in the sandier spots the purple heather and red sorrel are seen with most pleasing effect.

Dulwich Park of 72 acres, and Brockwell Park of 84 acres, are two of the most frequented of the open spaces in Camberwell. The beauty of Brockwell Park consists in its wildness, for there are long stretches of undulating lawn, dotted here and there with fine specimen forest trees. There is one feature peculiar to this park which requires mention, and that is the old garden. When the park was taken over, it was used as the kitchen garden. It is walled in on all four sides, but the walls are covered with roses and fruit-trees. Inside the walls the garden is laid out in the old-fashioned formal geometrical style, and it is a pleasing specimen of the kind found at many old castles and halls. Dulwich Park was a free gift from the Governors of Dulwich College, and the principal entrance is near the old College Chapel, and close by the famous picture-gallery. The American garden with its rhododendrons, azaleas, and roses is a great attraction in the season ; but the chief feature of the park is the rockery planted with showy Alpine and rock plants, which are said to be second only to those at Kew.

The Lake, Dulwich Park

We can make only a passing reference to Southwark Park of 63 acres, which is situated in Rotherhithe, and is a great boon to the people in this crowded district.

5. The River Thames. The Lea. The Ravensbourne. The Bridges and Tunnels.

In an early chapter we read that London was founded on a site about 60 miles from the coast, and we also learnt that "London" and "Thames" are the only Celtic words remaining in this area to remind us of the British occupation of our country. Now, as the Thames has played such an important part in the growth and development of London, it will be necessary to devote a little time to the study of this our greatest river. A recent writer has said that "The river has made London, and London has acknowledged its obligations to the Thames. It was the Silent Highway along which the chief traffic of the City passed during the Middle Ages.... The river continued to be the Silent Highway until the nineteenth century, when it lost its high position. With the construction of the Thames Embankment the river again took its proper place as the centre of London, but it did not again become its main artery."

The Thames, indeed, with its tides and its broad shining waters, has always been the source of London's wealth, and has been well named by one poet "Father Thames," and by another writer the "Parent of London." Throughout our history and literature the Thames plays a prominent part, and we shall find in the pages of this book many references to it. With our English poets it has been a favourite theme, and we find such expressions

as the "silver-streaming Thames" of frequent occurrence, while Denham has sung its praises in some noble
couplets :—

> "Oh could I flow like thee, and make thy stream
> My great example, as it is my theme :
> Though deep yet clear, though gentle yet not dull,
> Strong without rage, without o'erflowing full."

The watermen of London were long famous, and
many were the sports on the Thames that gave colour to
the life of Londoners.   There are many records of the
Thames being frozen over in severe winters, and some of
the Frost Fairs on the ice were of considerable duration.
One of London's historians of the sixteenth century gives
us some idea of the plentifulness of the fish caught in the
Thames of London.   "What should I speak," says
Harrison in 1586, "of the fat and sweet salmons, daily
taken in this stream, and that in such plentie as no river
in Europe is able to exceed it ? "   The first salmon of the
season was generally carried to the King's table by the
fishermen of the Thames.   A sturgeon caught below
London Bridge was carried to the table of the Lord
Mayor ; if above bridge, to the table of the King or Lord
High Admiral.

London has had great pageants on its river, and
in Stuart times the Dutch ships came on its stream
almost within gunshot of the Tower.   Queen Elizabeth
died at Richmond, and her body was brought in great
pomp by water to Whitehall.   Nelson's body, too, was
carried in great state by water from Greenwich to
Whitehall.   Many a state prisoner, committed from

Fair on the Thames, February 1814

the Council Chamber to the Tower, was taken by water; and we all remember that striking scene in our history, when the Seven Bishops were carried on the Thames to the Tower. Almost as a sequel to the last event, James II himself fled from London by water, and, in his flight, threw the Great Seal of England into the Thames.

Such, then, are a few of the historic landmarks that draw our attention to the river that has made London the capital of the British Empire.

It will be seen by a reference to the map that the Thames divides London into two unequal portions. It is navigable and tidal throughout its course through London; and from its source in the Cotswold Hills to the Nore the direct length is 120 miles, although with the windings it is probably about 220 miles in length.

The Thames, or Tamesis as it was once called, is the earliest British river mentioned in Roman history. Its name, as we have seen, is of Celtic origin, and its derivation is probably the same as that of the Tame, the Teme, and the Tamar in other parts of England. The upper part of the main stream is often called the Isis, and not the Thames, until it has received the waters of the Thame near Dorchester in Oxfordshire. In its upper course it passes through some of the finest agricultural country, while below London Bridge it is one of the most important commercial highways in the world. The Thames begins to feel the tide at Teddington, and from there to the Nore, a distance of 68½ miles, the tide ebbs and flows four times in the day. The force of the tide is very great and its power can be well seen at Blackfriars Bridge, where the water swirls round the piers, and rushes through the arches like a mill-race.

The Thames enters the County of London at Hammersmith, where it is crossed by a bridge. The western section of the river from this point to the Temple

Blackfriars Bridge

belongs to our second volume on London ; we will therefore begin our journey down the river from the City boundary to the point in Woolwich where it ceases to be a river of the County of London. It will be well, however, to bear in mind that the jurisdiction of the City of London, and of the Port of London, extends much further to the east. This will be made clear in a later chapter dealing with the newly-formed Port of London Authority which takes the place of the Thames Conservancy Board.

In passing hurriedly along the Thames, we shall notice that the northern or Middlesex side is much more interesting than the southern, or Surrey and Kent side. The Thames Embankment on the north is one of the greatest London improvements of the nineteenth century, and with its fine public buildings forms a striking contrast to the wharves and quays on the Surrey bank. In mid-stream, the river is crowded with coal-barges and every kind of river craft, while moored close to the Embankment is the *Buzzard*, a fine ship used by the Naval Territorials. When we near Blackfriars Bridge, which has been recently widened, we get a fine view of St Paul's and Bow Church, and realise how the former dominates the City.

We now reach Queenhithe, the old City landing-place, which was so famous in Plantagenet days. On the Surrey side, Southwark Cathedral and St Olave's are very prominent and recall the former glories of the Southwark and the Bankside of Shakespearean times.

Our boat now carries us to London Bridge, the very beginning of London, for here the Thames was first spanned by the Romans. The present stone bridge is a grand piece of work, and the five arches take the place of the former twenty-one, which acted as a kind of weir. Billingsgate, the fish-market of London, is next noticed, and then the Custom House Quay with its pigeons flying

The Pool and London Bridge

overhead comes in view. The Custom House is a hand-some building with fine columns on the river front.

We are now in the "Pool," or that part of the Thames which begins just below London Bridge. This is in many ways the most important part of the river, and the estuary is crowded with ships as far as the eye can reach. The Tower, square and massive, is on our left, and we soon pass under the Tower Bridge, the last or most seaward of the City bridges. A little below the Tower are St Katherine's Docks, enclosed by ware-houses, and these are succeeded by the London Docks on the north, and by the Surrey Commercial Docks on the south side. A little below the Pool, where the river takes an abrupt bend in its course at Limehouse Reach, is one of the entrances to the West India Docks, and further south the entrance to the Millwall Docks. The West India Docks run right across the tongue of land called the Isle of Dogs, and open into Blackwall Reach. As we pass down Limehouse Reach we notice Deptford on the right. Here there was a Government Dockyard until 1869, but now it is noteworthy only for the Royal Victualling Yard, and its Foreign Cattle Market.

Leaving Deptford, Greenwich Hospital comes into view, and this monument to the genius of Wren is seen to great advantage from the river. Still further down the Thames is Woolwich Arsenal, the largest Government ordnance depôt. The river below Woolwich, and nearly all the way to its mouth, lies between marshes, over which the ships appear to be sailing across the grass, as in a Dutch picture. Opposite Greenwich on the north

side are the East India Docks at Blackwall, and here
it is that the Thames on the north side terminates.

There are two tributaries of the Thames which may
be noticed in this section, before we consider the bridges
and tunnels of the river.

The Lea, or Lee, as it is sometimes written, forms the
boundary river of London on the east, from a point in
Clapton to its entry into the Thames at Blackwall opposite Greenwich marshes. There is every reason to
suppose that, from Clapton onward to its junction with
the Thames, the Lea once had a very considerable estuary,
and ancient Danish boats and other remains have been
dug up in recent years in the neighbouring marsh-lands.
The valley of the Lea in its lower course was the scene of
many a hard-fought struggle with the Danes, and it is a
matter of common history that Alfred diverted the stream
of the river, and left the Danish boats high and dry at
Ware, so that his enemies had to escape overland. At
the present time, the lower portion of the Lea is marked
by a series of great reservoirs of the Metropolitan Water
Board, and these supply a population of two million
people with water. Large tracts of marsh-land still
remain on either side of the lower Lea and formed in
ancient times a sure defence of the City on the east.
Now roads and railroads cross the Lea, where not so long
ago were only fords.

The most easterly stream flowing into the Thames on
the south side is the Ravensbourne. It rises near Keston
Common in Kent, and flows for a distance of ten miles
past Bromley and Lewisham to Deptford at its mouth.

The Ravensbourne, in its higher course, was once a beautiful river, but now it flows through a very poor and populous neighbourhood.

The Thames as it flows through the County of London is crossed by a number of fine bridges, from Hammersmith Bridge in the west to the Tower Bridge in the east. And yet, not two centuries ago, London Bridge was the only bridge over the Thames in London. It is still the most important, for it connects the City, the centre of London's business, with Southwark on the Surrey side of the river. Besides London Bridge in the eastern portion of the County of London, there are Blackfriars Bridge, Southwark Bridge, and the Tower Bridge.

On account of its antiquity and present importance, London Bridge must first claim our attention. The Romans probably built the first bridge over the Thames, and it has been remarked that the two chief events in the history of Roman London, the building of the bridge and the building of the wall, are the only two events of the period which had any permanent effect on its later existence. But for the wall, and but for the bridge, Roman London might as well never have been built. This bridge over the Thames was required by the Roman roads, and it is conjectured that Claudius crossed the river by the Roman bridge. Most historians assign the building of London Bridge by the Romans to the year A.D. 78 ; but with regard to this and other kindred matters we shall make reference in a later chapter. Old London Bridge crossed the river just east of the present bridge.

London Bridge, 1757

For a long time it was a wooden bridge with a fortified
gate.   There is no doubt that the Saxons built wooden
bridges over the Thames near the site of the present
bridge, but they were carried away by floods or destroyed
by fire.   At length, in 1176, Henry II instructed Peter,
the curate of St Mary Colechurch, to construct a stone
bridge at this point, but the work was not completed till
1207.   A chapel, dedicated to St Thomas of Canterbury,
was built upon the bridge, and a row of houses sprang up
on each side, so that the bridge resembled a continuous

London Bridge, 1784

street.   It was terminated on both banks by fortified
gates, and on their pinnacles the heads of traitors were
exposed.   This bridge was one of the chief sights of
London, and poets sang its praises.   It was neverthe-
less very inconvenient, for its narrow arches impeded
navigation, acting as a kind of weir.   The houses on each
side made the passage narrow and dark.   They were
removed in 1757, and the bridge altogether demolished in
1832.   The present bridge, designed by John Rennie,
was opened in 1831, and was widened in 1904.   It has
five granite arches, of which the central one has a span of

152 feet, and the lamp-posts on the bridge are cast from
the metal of French cannon captured in the Peninsular
War.   London Bridge divides London into "above" and
"below" bridge, and as we look down the river we see
the Port of London.   To the Pool—the part immediately
below the bridge—sea-going vessels of the largest size
have access.

Blackfriars Bridge is a continuation of Farringdon
Road, which it connects with Blackfriars Road on the
Surrey side.   It was built by Cubitt, and opened in 1869.
The present bridge of five arches occupies the site of a
stone bridge, the piers of which had given way.   It is an
iron structure, supported by granite piers, and from it an
excellent view of St Paul's and some of the City churches
may be obtained.   In 1907 it was found necessary to
widen this bridge, and on the completion of the work, in
1909, tramway lines were carried over it from the Victoria
Embankment to the Surrey side.

Southwark Bridge is one of Rennie's bridges, and was
opened in 1819.   It has three iron arches supported by
stone piers, but the traffic is not considerable owing to the
approaches.   This bridge connects Upper Thames Street
in the City with Southwark Bridge Road on the Surrey
side.

The Tower Bridge is the last and most attractive of
the London bridges.   It was designed by Sir Horace
Jones and Mr Wolfe Barry, and cost over £1,000,000.
The money was provided by the City Corporation, and
the bridge was opened in 1894.   A bridge here was much
needed to connect Bermondsey with the Minories, and

The Tower Bridge

also to relieve the congested traffic of London Bridge. The difficulty was to construct a bridge below the Pool that should give access to the wharves and markets, and it was surmounted by the present remarkable structure, which has a permanent footway 142 feet and a carriage way 29½ feet above high-water mark. The central span of the latter is 200 feet long and is fitted with two bascules or drawbridges which can be raised in the space of two minutes to admit of the passage of large vessels. When the bascules are raised, access to the high footway is gained by means of lifts and stairs in the supporting towers, which are raised on piers from the river-bed to a height of 246 feet, and are connected with the river-banks by permanent spans of 270 feet, suspended on massive chains hanging between the central towers and smaller castellated towers on shore. The framework of the bridge, including the central towers, which are cased in stone, is of steel. Large electric bells are fixed at the entrance to each approach, and they are rung when ships are passing through, so that drivers may be warned when the bascules are being raised. The bascules are worked by hydraulic power, by means of a powerful counterpoise. A 21-inch steel bar, which bears the weight of 1000 tons, is the pivot on which the whole works.

Besides the bridges over the Thames there are several tunnels beneath its waters facilitating communication between the opposite shores. The Thames Tunnel, two miles below London Bridge, connects Wapping on the left bank with Rotherhithe on the right. It was planned by Brunel in 1843, and carried to completion after great difficulties. It is not used for passenger traffic, for it was sold in 1865 to the East London Railway Company to be used in connecting the Great Eastern and North London Railways with those south of the Thames.

The Blackwall Tunnel, opened in 1897, goes from Blackwall on the north side to Greenwich on the south. It is about 1¼ miles in length, and is the greatest engineering work of the kind ever constructed. The making of the tunnel cost over 1¼ millions sterling, and it is a most important means of communication, being available for traffic of all kinds. The chief feature as an engineering work was in boring under the river, which was accomplished by means of a shield forced forward by hydraulic pressure, the excavation being carried out under compressed air.

Greenwich Tunnel, opened in 1902, is for pedestrians only, electric lifts conveying passengers to and from the tunnel. It connects Greenwich with Poplar and is 1217 feet in length.

Rotherhithe Tunnel, opened in 1908, gives communication between Rotherhithe and the north side. It is a most important highway for both vehicular and pedestrian traffic, and has cost about £1,750,000.

In addition to the bridges and the tunnels, the Woolwich Free Ferry connects Woolwich on the south with North Woolwich on the north, or Essex side. We may realise its value when we learn that it is used by about 6,000,000 people during the year.

6. Rivers of the Past—The Fleet and the Walbrook.

We may now proceed to consider the rivers and streams that once fell into the Thames. The most important of those on the north were the Westbourne, the Tybourne or Tyburn, the Holebourne or Fleet or Wells River, and the Walbrook. Not one of these streams now runs above ground, and if they flow, it is merely as underground sewers. Their courses, however, can still be traced by the names of places that formerly stood on their banks. As regards the streams of the south, they have little bearing upon the history of ancient London. The chief fact to bear in mind about the south is the vast extent of marsh-land in ancient days, now covered with thousands of houses. A few streams crossed the marshes, and we may mention the Falcon Brook and the Effra. For the present, we will consider the Fleet and the Walbrook.

The Fleet River was also known as the Holebourne, the Wells River, and the Turnmill Brook. Its most ancient name was the Holebourne, i.e. the stream that flows in a hollow. It was named the Wells River or the River of Wells because of the great number of wells or springs that fed it; and it received the later and more familiar name, the Fleet, because at its mouth it was a "fleet" or channel covered with shallow water at high tide. It was a very considerable stream formed by two branches, one of which rose in the Vale of Health,

Hampstead, and the other in Ken Wood between Highgate and Hampstead. These branches joined at a place in Kentish Town Road, and the river then flowed down the present Farringdon Road, Farringdon Street,

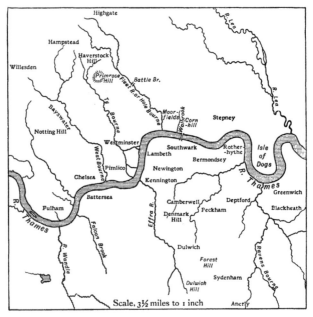

The Streams of Ancient London

and New Bridge Street into the Thames. Some of the wells that gave the river one of its names were Clerkenwell, Godwell, and Bridewell. For many years the Fleet was the western boundary of London, and as it afforded a water-way for some distance inland it was of much value for trading purposes. In later times, however, it became an open noisome sewer, and its banks were cumbered with the rudest houses until the Great Fire swept all away. After this a stone embankment on either side of the Fleet enclosed its waters, and the lower part of the river became a canal 40 feet wide, with wharves on both sides. Four bridges were built over the canal—at Bridewell, Fleet Street, Fleet Lane, and Holborn. The canal was not a success, for the stream again became choked, and once more returned to its sewer-like condition. As a result, the part between Holborn and Fleet Street was covered over in 1737, and the lower part between Fleet Street and the Thames in 1765. The Fleet Ditch, as it was sometimes called, is often mentioned in the annals of London, and its name survives in the present Fleet Street, one of the busiest thoroughfares of the City.

The Walbrook was formed by the junction of two streams which rose in Hoxton and in Moorfields. It entered the City through a culvert a little to the west of London Wall, crossed Lothbury, and passed under the present Bank of England, and thence across the Poultry. It did not run down the street now called Walbrook, but on the west side of it, and so made its way to the Thames between Friars' Alley and Joiners' Hall. The outfall has been changed, and the stream now runs under Walbrook, and finds its way to the Thames at Dowgate Wharf. The banks of the Walbrook were evidently a favourite place for the villas of the wealthy Romans, for many Roman remains have been found along its course.

It is worth noting that two low hills stood on either side of the Walbrook, and it was on the western "dun" that London was first built. The bed of the Walbrook has been frequently found in recent City excavations. It was crossed by many bridges, and probably afforded the first water-supply of London. It was an important factor in the division of the City into wards and also in the laying out of the streets.

7. The Water=Supply of London—Past and Present.

The water-supply of a great city is of the utmost importance, for on its good quality and constant flow depend largely the health and happiness of its people. In considering the supply of water to London we must remember that ancient London and the many parishes now comprised in the modern county arose on sites where a supply of good drinking-water could readily be obtained from natural springs and brooks, or by means of wells. The earlier settlements were made on the tracts of gravel and sand, and the growth of London was regulated for a long period by the distribution of these water-bearing strata. Thus we find that the City expanded westwards to Chelsea, Kensington, and Hammersmith; southwards to Clapham and Camberwell; eastwards to Bow and Hackney; and northwards to Islington. Such districts as Camden Town, Kentish Town, and Kilburn were not populated until a supply of drinking-water from a distance was brought in conduits.

It is a matter of history that, from 1680 to 1840, some of the London wells and springs attained fame as holy wells and spas. Thus we read of Beulah Spa, Bermondsey Spa, Islington Spa, Holywell, Clerkenwell, and Sadler's Wells, while at Well Walk, Hampstead, a chalybeate spring was utilised until quite recently. The first conduit for the supply of water to London was that of Tyburn, which was completed in 1239, when water was conveyed in leaden pipes to the City. Much water was also obtained in buckets from the river, and in 1582 the supply was facilitated by means of water-wheels attached to the arches of old London Bridge. After a while wooden conduits were used, and a more extended system of supply to houses was introduced. In opening some of the London streets it is no uncommon thing to find these wooden conduits as they were placed a long time ago. Some of these relics of the early times of London's water-supply may be seen in various museums.

With the growth of London, the supplies of water from the gravel soils became contaminated, and the water of the Thames near London Bridge was very bad. From the close of the seventeenth century and onwards to 1855, companies were formed for taking water from the Thames near Charing Cross, and higher up; but since that year, no water has been drawn by any company from the Thames below Teddington Lock.

Sir Hugh Myddelton was the pioneer in bringing water to London from a distance. In 1608, he commenced the cutting of the New River and five years later this artificial channel was completed. As a result

of his efforts the New River Company was formed in
1619, and down to the present century it has been of
the greatest service to London, supplying an abundant
quantity of excellent water from the River Lea and
from springs in the Chalk, as well as from deep wells
sunk into the Chalk.

The New River, Clissold Park

Many artesian wells have been sunk through the
London Clay into the Lower Tertiaries and Chalk, and,
since 1790, breweries and other large establishments have
used this means of obtaining their water. One of the
deepest borings through the Chalk in the London Basin
is that at Kentish Town, where various beds have been
passed through to a depth of over 1300 feet.

The water-supply of London and the surrounding districts is now controlled by the Metropolitan Water Board. Before the year 1902, this great area was supplied by eight London Water Companies; but by an Act passed in that year the Water Board was created for the purpose of purchasing and managing the undertakings of those companies. The Metropolitan Water Board area is much greater than that of the County of London and extends from Ware in Hertfordshire to Esher in Surrey, and from Romford in Essex to Chevening in Kent.

The Metropolitan Water Board in 1908 had to supply a population of more than 7,000,000 persons, and to deliver a daily average of 224,000,000 gallons. The whole of this water is obtained from the Thames and the Lea, and from various springs and wells in the locality. The water mains have a total length of 6041 miles, and the water they bring to the vast population of London and its environs is of a very high standard of excellence and purity.

The great fault of the water supplied over this area is "hardness," that is, that it contains a quantity of bicarbonate of lime, yet it is well known that many of the healthiest districts are those with hard water. It is this hardness of chalk waters which furs our kettles and wastes our soap; but it is not easy in the London area to obtain other than hard water, for much of it is derived from wells in the Chalk.

8. Geology.

In geological language, London is said to be situated in a "basin"—the "London Basin." This basin has been carved out of the System belonging to the early Tertiary period which is called Eocene. The solid foundation is composed of the Chalk, a Secondary formation here about 600 feet thick. This it is which really constitutes the London Basin, whose broad rim comes to the surface in the Chiltern Hills in the north and north-west, and in the North Downs on the south. Although the Chalk may be called the basement-rock of the London Basin, yet wells have been sunk in the middle of that area to so great a depth as to pass through it into the beds below.

For the purpose of this chapter we may commence our description of the rocks under London with the Gault, the formation that occurs almost universally in the London Basin below the Chalk. The Gault is a marine deposit, and seems to have been deposited in a moderately deep, quiet sea, and not along a shore-line. It is a bluish clay and varies in thickness from 130 to 200 feet at such deep borings as those at Tottenham Court Road, Kentish Town, and Mile End.

Above the Gault is the Upper Greensand, which is also of marine origin. It consists of clayey green sand varying in thickness from 10 to 30 feet. In all cases, it has been found in the same deep borings as the Gault.

The Chalk comes after the Gault, and is perhaps the best known rock in England. It is a soft white limestone, but contains beds of a harder nature. In the London Basin, except in the south-east corner, the Chalk with flints alone occurs at the surface. In deep borings through the Chalk in London, the thickness of this formation varies from 645 to 671 feet. By its fossils it is proved to be the deposit of a fairly deep sea, something of the same character as that now forming in the mid-Atlantic Ocean. The Chalk is not much exposed in the County of London, but there are some pits near Deptford and near Lewisham where it may be seen.

Over the Chalk come a number of thin but varying beds known as the Lower London Tertiaries. To each of these divisions a local name has been given. The Thanet Beds are the lowest, and are so named from the fact that they are the only Tertiary formation in the Isle of Thanet. The Woolwich and Reading Beds succeed, and are named from their occurrence in the neighbourhood of those places. The upper formation of Oldhaven Beds follows the last, receiving its name from the good section of it which may be seen at Oldhaven Gap, near the Reculvers in Kent.

The Thanet Beds consist almost wholly of fine soft sand, very pale grey or buff, and very compact. They are without either pebbles or fossils and have a thickness of about 30 feet. Sections rarely reach the surface in the County of London, but the Thanet sand may be well seen at Plumstead, Woolwich, and Lewisham.

The Woolwich and Reading Beds are a group of clays and sands having a thickness of 60 feet or less. Shells of an estuarine character are found in the clay and this fact proves that the beds were deposited at or near the mouths of streams. Sections of these beds may be seen at Woolwich, Charlton, and Lewisham.

The Oldhaven and Blackheath Beds consist mostly of a bed of perfectly-rolled flint pebbles, in a base of fine, sharp, light-coloured sand. The thickness is as much as 50 feet, and fossils are met with in parts. Interesting open sections of the Blackheath Beds may be seen at

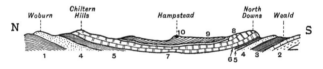

The London Basin

Eltham, Bostall Heath, Woolwich, Plumstead Common, and Blackheath.

Overlying the last beds is a great mass of clay known as London Clay. Although this formation takes its name from the metropolis, it is well to remember that it extends from Marlborough on the west to Yarmouth on the north-east. In the neighbourhood of London the Clay is 400 feet thick, and many fossils have been found in it. The London Clay forms a very broad band right through the London area from south-west to north-east, and excellent sections may be seen at Plumstead Common, Hampstead, and Highgate.

Above the London Clay come a group of sands which may be comprised under the name Bagshot Beds. As a whole they form a more or less barren sandy tract of rising ground, which is partly open, but sometimes covered with fir and larch. The hills of Hampstead and Highgate are perhaps the most prominent heights in north London, and although the Bagshot Beds cap these hills, it must be remembered that their longer slopes are of London Clay.

In the London Basin, after the Bagshot Beds, we come to a great gap in the series of geological formations. The beds just named are Eocene, and we find nothing more till we are almost out of the Pliocene Period. Gravels, sands, and clays are found at various levels down nearly to the present level of the Thames, and this newer set of deposits may be classed under the term Drift. The Boulder Clay is one of the most important of these deposits. It is stiff and tenacious, and often studded with pieces of Chalk. Good sections of the Boulder Clay are rare but it can sometimes be seen in temporary openings and in roadside sections and ponds.

Passing over some deposits known as Brick earth, and Valley or River Gravel, we come to the last and newest deposits of the district. These Alluvial Deposits are confined to the very bottoms of the valleys in which rivers run. They comprise the strip of marsh-land or Alluvium which fringes the river over small areas above London, and over broader tracts in southern Essex and northern Kent. The Alluvial Deposits are from 12 to 20 feet thick, and the old river mud often contains bones of the ox, deer, and elk, as well as implements of stone, bronze, and iron.

It is advisable that, while reading this chapter, constant reference should be made to the geological map at the end of this volume. The reader is also advised to pay a visit to the Geological Museum at Jermyn Street, where there is a large model of London and the neighbourhood. It is on the scale of 6 inches to the mile, and owing to its great size is in nine sections. It represents an area of 165 square miles and gives an excellent idea of the geological structure of London.

9. Natural History.

Among all the English counties, London has least to attract the lover of natural history. Almost its entire area is given over to bricks and mortar, and outside the parks and open spaces there are few places where the flora and fauna can be studied in the same way as in the neighbouring counties of Essex, Kent, and Surrey. There are tracts in the north and in the south-east and south-west where the population is not so dense as in central London, but even in those districts streets are being made every year, with the consequent destruction of plant and animal life.

It would almost be easier to write about the trees and flowers in London many years ago than of those at the present time. London was once famous for its trees and

flowers. Vinegar Yard, Covent Garden, was the vineyard
of Covent Garden. Saffron Hill was once covered with
saffron. The red and white roses of York and Lancaster
were plucked in the Temple Gardens, and Daniel the

The Pines, Bostall Wood, Plumstead

poet, in the reign of Elizabeth, had an excellent garden
in Old Street, St Luke's. Gerard the herbalist in the
same reign had a choice assemblage of botanical specimens
in his garden at Holborn.

The flora of the south-east and south-west of the county of London is similar to that of Kent and Surrey. The blue wood-anemone (*A. apennina*) was formerly abundant as an introduced plant in Wimbledon Park, but is now extinct. The cowbane (*Cicuta virosa*) formerly grew by the Thames at Battersea. Gerard (1633) records that this plant grew in Moor Park, Chelsea, but of course it is no longer found there. The sea-aster (*A. Tripolium*) grew by the Thames near Battersea, but it, too, is no longer a plant of the county.

Furze, broom, briars, bracken, and heath are abundant on Barnes Common, and on Putney Heath and Wimbledon Common there grows scrub of stunted oak, of hazel, birch, and sallows, with plenty of tall furze. Wandsworth Common now grows nothing unusual, but formerly its speciality was the water-soldier (*Stratiotes aloides*), while Streatham Common was famous for *Senecio viscosus*.

The Plumstead Marshes and the flats below Woolwich and towards Erith have now been drained and put under pasturage. Aquatic plants, both rare and ordinary, grew here in great variety and abundance, but they have now almost disappeared. In the north of the county Hampstead Heath is the chief open space. The ground is there broken into pits and hillocks, and much bracken grows, with a few white and black thorns.

There is one survival in London which is worth a passing notice. Adjoining the Chelsea Embankment is the Physic Garden founded in 1673 and presented by Sir Hans Sloane in 1722 to the Society of Apothecaries on condition that 50 new varieties of plants grown in it

should be annually furnished to the Royal Society, until
the number so presented amounted to 2000. It was
famed for its fine cedars, the first to be grown in
England. They are now no longer existent, the last
having died in 1904. The great botanist Linnaeus visited
this garden in 1736, and Kalm the Swedish naturalist
in 1748. Towards the end of the last century, the
Apothecaries Society being no longer desirous of maintaining the garden, it was vested in the London Parochial
Charities in 1899. A committee of management was
appointed, and as a result of this change many improvements have been made. The plant houses have been
rebuilt, and a well-fitted laboratory and lecture room
erected. The place is now used by students of the
Royal College of Science, and members of various schools
and polytechnics. Courses of advanced lectures in Botany
are arranged by the University of London; and specimens
are supplied to the principal teaching and examining bodies
in the metropolis.

It is very noticeable when houses are cleared away
in London, and the space remains derelict for a time,
that all kinds of plants will grow and flourish. In 1909
there was a large tract of waste ground on the north of
the Strand, and in the month of July the area was a
blaze of purple. Amid the mass of vegetation that had
sprung up there were elder, cherry, loosestrife, a few plants
of willow herb, a great crop of rape, and here and there
a thistle. In a short space of time it was not difficult to
find as many as twenty different species, and the question
arose how they came to grow in such an unlikely spot.

Air-borne seeds, such as dandelion, settled in this area and grew with little difficulty. Other seeds were carried to Aldwych in the hay or chaff given to horses at the time when the buildings were being demolished and the débris carted away, while the elder and cherry were probably brought by birds.

The parks and open spaces of London are the haunt each year of a far larger variety of wild birds than is generally supposed. It is only during the great migration times in spring and early autumn that a full idea can be obtained of the large number of species which pause for a few hours in the chief London parks as they seek or leave their summer nesting-places. Among these passing visitors to London are the wheatear, the redstart, the sandpiper, the kingfisher, and the great crested grebe. The wheatear and the brilliant kingfisher are by no means unknown in Hyde Park and Kensington Gardens.

Besides those birds which, in London, are essentially birds of passage, there are many birds which are regular residents for the whole or part of the year, and thus save London from the reproach of being a birdless county. There are some, indeed, which find greater security in London than in the country. One of the most conspicuous examples of this class is the carrion crow, which is common in some of the parks and squares, more especially in Kensington Gardens. The crow in London maintains a steady hostility to rooks, and the Gray's Inn rookery has been so harried of late years by the crows, that this famous colony is gradually diminishing and disappearing. In 1836 there were 100 nests of rooks in Kensington Gardens, but owing to the terrorism of the crows, the rookery there is well-nigh extinct. The brown owl is another bird of prey which seems glad to find refuge in London. It is not uncommon among the

Sea Gulls on the Embankment

old trees in the larger London gardens, planted a century or more ago, and it has adapted itself to the London life of to-day.

The sparrow is looked upon as a London bird and has every equipment for the needs of London life. He is there viewed with a tolerance, and even with a sentimental affection, which is not extended to him in the country. In London he is ubiquitous, and seems to find a satisfaction in placing his nest in the most ridiculous positions. The song-thrush and blackbird often sing more vigorously in London during the winter than in most places in the open country. Swallows, martins, and sand-martins are sometimes seen in considerable numbers on the Round Pond, and on the Serpentine, especially in cold and frosty Aprils. The spotted fly-catcher attempts to nest every year in Hyde Park and Kensington Gardens, and the reed-warbler occurs on the London list, for it sometimes pauses on migration in the thickest of the reeds at the head of the Serpentine.

Perhaps the most striking example of a recent addition to the birds of London is the annual winter visit of the black-headed gulls, which haunt the Thames and the ponds in the parks from October until March. The flocks include from time to time a few gulls of other kinds, the herring gulls being commonest. The gulls first visited London in large numbers in the hard frost of 1895, and have never since abandoned it. Still more remarkable, however, is the introduction of that most wary of all country birds, the wood-pigeon. Unknown a few years ago not only in London but in its near neighbourhood, they became established about 1895 and have remarkably increased. They have become extremely tame and may be seen feeding in dozens in the parks, and even in the roads.

The ornamental waters in the parks are so well

Canonbury Tower, one of Queen Elizabeth's hunting seats

stocked with different breeds of ducks that it is impossible
to say to what extent they are frequented by genuine
wild-fowl. There is no doubt, however, that such
visitants are considerable. There are upwards of 400
wild-fowl in the splendid collection at St James's Park
which breed on Duck Island. The herons, whose wings
are clipped, have been there since 1908, and besides black
and white swans there are many sorts of geese, wigeon,
teal, and mallard.

There was a time when the kite was as familiar in
London streets as the sparrow is now. The swifts used
to circle and glide over what is now the densest part of
the City. Woodcocks were shot in Piccadilly not a
hundred years ago. Although many changes have taken,
and must take, place in the bird life of London, there is
no sign that the great city will ever lose the interest
afforded by the presence of wild birds.

10. Climate and Rainfall. Greenwich Observatory and its Work.

The climate of a country or district is, briefly, the
average weather of that country or district, and it depends
upon various factors, all mutually interacting; upon the
latitude, the temperature, the direction and strength of
the winds, the rainfall, the character of the soil, and the
proximity of the district to the sea.

The differences in the climates of the world depend
mainly upon latitude, but a scarcely less important

factor is proximity to the sea. Along any great climatic zone there will be found variations in proportion to this proximity, the extremes being "continental" climates in the centres of continents far from the oceans, and "insular" climates in small tracts surrounded by sea. Continental climates show great differences in seasonal temperatures, the winters tending to be unusually cold and the summers unusually warm, while the climate of insular tracts is characterised by equableness and also by greater dampness. Great Britain possesses, by reason of its position, a temperate insular climate, but its average annual temperature is much higher than could be expected from its latitude. The prevalent south-westerly winds cause a movement of the surface-waters of the Atlantic towards our shores, and this warm-water current, which we know as the Gulf Stream, is one of the chief causes of the mildness of our winters.

Most of our weather comes to us from the Atlantic. It would be impossible here within the limits of a short chapter to discuss fully the causes which affect or control weather changes. It must suffice to say that the conditions are in the main either cyclonic or anticyclonic, which terms may be best explained, perhaps, by comparing the air currents to a stream of water. In a stream a chain of eddies may often be seen fringing the more steadily-moving central water. Regarding the general north-easterly moving air from the Atlantic as such a stream, a chain of eddies may be developed in a belt parallel with its general direction. This belt of eddies or cyclones, as they are termed, tends to shift its position, sometimes

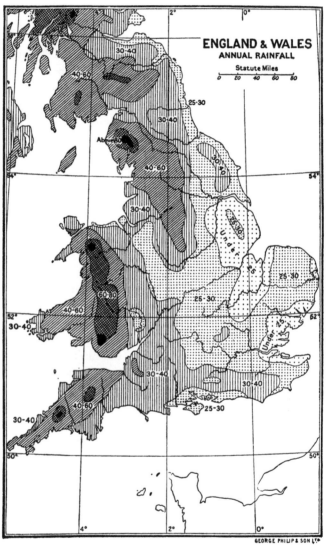

(The figures give the approximate annual rainfall in inches)

passing over our islands, sometimes to the north or south of them, and it is to this shifting that most of our weather changes are due. Cyclonic conditions are associated with a greater or less amount of atmospheric disturbance ; anticyclonic with calms.

The prevalent Atlantic winds largely affect our island in another way, namely in its rainfall. The air, heavily laden with moisture from its passage over the ocean, meets with elevated land-tracts directly it reaches our shores—the moorland of Devon and Cornwall, the Welsh mountains, or the fells of Cumberland and Westmorland —and blowing up the rising land-surface, parts with this moisture as rain. To how great an extent this occurs is best seen by reference to the accompanying map of the annual rainfall of England, where it will at once be noticed that the heaviest fall is in the west, and that it decreases with remarkable regularity until the least fall is reached on our eastern shores. Thus in 1906, the maximum rainfall for the year occurred at Glaslyn in the Snowdon district, Carnarvonshire, where over 205 inches of rain fell ; and the lowest was at Boyton, Suffolk, with a record of under 20 inches. These western highlands, therefore, may not inaptly be compared to an umbrella, sheltering the country further eastward from the rain.

The above causes, then, are those mainly concerned in influencing the weather, but there are other and more local factors which often affect greatly the climate of a place, such, for example, as configuration, position, and soil. The shelter of a range of hills, a southern aspect, a sandy soil, will thus produce conditions which may differ greatly from those of a place—perhaps at no great distance—situated on a wind-swept northern slope with a cold clay soil. The character of the climate of a country or district influences, as everyone knows, both the cultivation of the soil and the products which it yields, and thus indirectly as well as directly, exercises a profound effect upon Man.

In considering the climate of the county of London we must bear in mind that it is not a maritime county like Essex, and so has not the modifying influence of the sea. It will also be well to remember that, in point of size, London is the smallest of our English counties, and therefore we must not expect to find the variations in its climate so noticeable as those in Kent or Essex.

It is of the greatest importance to have accurate information as to the prevailing winds, the temperature, and the rainfall of a district, for the climate of a county has considerable influence on its productions, its trades and industries, and its commerce. Our knowledge of the weather is much more definite than it was formerly, and every day our newspapers contain a great deal of information on this subject. In London there is the Meteorological Society, which collects information from all parts of the British Isles relating to the temperature of the air, the hours of sunshine, the rainfall, and the direction of the winds.

The British Isles and coasts have been divided for these purposes into 12 main districts, and the newspapers daily publish the forecasts issued by the Meteorological Society of the probable weather for the twenty-four hours next ensuing, ending at midnight. Thus, for September 30, 1909, the following was the forecast for London, which is placed in the South-east England district :—"Calms and very light variable breezes, north-easterly to north-westerly; dull to fair or fine; local rain and mist; cool." Warnings are also issued when necessary, so that certain districts may be prepared for the rough weather that may be expected. Besides this information, some of the newspapers print maps and charts to convey the weather intelligence in a more graphic manner.

A glance at a map of the world will show that, as the British Isles are in the same latitude as Central Russia, Southern Siberia, Kamtchatka, and Labrador, they get the same amount of heat from the sun and the same duration of day and night, summer and winter ; but, with us, the direction of the prevailing winds renders available throughout the year much of the heat which the sun has radiated on more southern regions.

The prevailing winds of London, like those of the British Isles generally, are south-westerly. Indeed the wind blows from the south-west for a greater number of days in each month than from all other directions together. A knowledge of this fact helps us to under-stand that the west end of London is the least smoky and therefore the best quarter for residence. For a short period of the year, London suffers from the east wind, and during its prevalence in March and April, the air is dry and catarrhal complaints are common.

London occasionally suffers during the two or three winter months from thick fogs which hang over the

city like a black pall. Their density is mainly due to
coal-smoke, and many efforts have been vainly made to
remedy them. Fogs can nowhere be avoided in the
London area, though they are less dense at Hampstead
and Highgate in the north, or at Streatham in the south,
than at Whitechapel or Rotherhithe. There is even a
Smoke Abatement Society, but up to the present there
has been little or no alleviation of the annoyance.
Besides the health point of view, there is the commercial
aspect of these London fogs to be considered, for the
dislocation of business is enormous, while the extra ex-
penditure caused by resort to artificial light is a great tax
on tradesmen and others. In London, the yellow fogs are
humorously known as "London Particular," and Dickens
often refers to them in his works by this name. London
has also its mists, and these occur throughout the year.
To the artistic eye, a London mist is really beautiful,
and Whistler has taught people to appreciate these mists
in relation to the river by his "Nocturnes."

The warmest month in London is generally July,
when an average temperature of 64° Fahr. prevails.
January, the coldest month, has an average temperature
of 39°. The mean temperature of England was 48·7°
in 1906, and that of London was 50·7°. With regard to
the hours of bright sunshine in the same year, we find
that while London had 1734·5 out of a possible total of
4459, the average for all England was 1535·5, so that
London was much above the country as a whole.

The statistics with regard to the rainfall are arranged
in an annual known as *British Rainfall*, and from it we

can find exactly recorded the number of inches of rain that fall at about 4000 stations throughout the British Isles. In the County of London there are many observers who keep one or more rain-gauges, and register the results obtained. Every year these facts are tabulated and forwarded to the editor of *British Rainfall*. In the County of London there are two very important stations for recording meteorological statistics. The first is Greenwich Observatory, which is a Government establishment, and the second is at Camden Square, where Dr H. R. Mill, the editor of *British Rainfall*, has a station.

Now let us look at the rainfall statistics at Greenwich for 1907. During that year there were 163 rainy days with a total rainfall of 22·25 inches. The two wettest months were April and October, each having a rainfall of over 3 inches, while March, July, and September were the driest with a rainfall of less than one inch. If we turn to the records at Camden Square, which is in the north of London, we find that rain fell for 418·8 hours on 175 days, giving a total rainfall of 23·01 inches, which is rather higher than that of Greenwich. The average rainfall of 50 years at Camden Square is 25·07 inches; the lowest rainfall of 17·69 inches was in 1898, and the highest of 38·0 inches was in 1903. For purposes of comparison it may be mentioned that the rainy days for England and Wales were 203 in 1907, and for the same year the rainfall was 36·11 inches. It will thus be seen that London is far below the rest of the country. This is largely due to the fact that the rainfall of England, as already stated, decreases generally as we travel from west to east (p. 67). The highest rainfall in 1907 was at Llyn Llydaw Copper Mine,

Greenwich Observatory

Carnarvonshire, where no less than 196·16 inches were measured. The least rain in 1907 fell at Clacton-on-Sea, where the record was 16·66 inches. Now let us compare these two extremes with those of London. The highest rainfall registered in London in 1907 was a total of 26·4 inches at Brockwell Park, Lambeth, and the lowest of 17·88 inches was at Battersea, Nine Elms.

Now to summarise the main facts with regard to the climate of London, we may say generally that it is dry, with a rainfall far below the average of England and Wales. The climate is healthy and the prevailing winds are from the south-west. The drawbacks to the climate are the fogs of November and December, and the biting east winds of the spring. Perhaps the best testimony to the healthiness of London's climate may be gathered from the vital statistics for 1908. In that year the death-rate per 1000 was only 13·8 against 15·2 for the whole country.

Greenwich Observatory is world-famous. It was founded by Charles II in 1675, and designed by Sir Christopher Wren. For the purposes of navigation the staple work of the Observatory has always been the observation of positions of the moon and fixed stars, to which has been added the care of the chronometers used in the Royal Navy. The meridian marked out by the Transit Circle, the instrument with which these observations are made, is the zero of longitude used by most of the nations of the world, and the mean time of the meridian of Greenwich is the legal standard time for Great Britain. At 10 o'clock and 1 o'clock each day, the accurate Greenwich time is telegraphed to the General Post Office for distribution over the whole country. The observations of positions of the heavenly bodies have been supplemented by the addition of branches dealing with meteorology, magnetism, the observation of the solar surface, and

The 30-inch Reflector at Greenwich

celestial photography. The photographs of comets, nebulae, and the small satellites of planets, taken with the 30-inch Reflector, compare favourably with those taken in other parts of the world under better climatic conditions. The Astronomer Royal lives at Flamsteed House, which forms the main part of Wren's building. The curiously-shaped domes covering the telescopes, the largest of which is a refracting telescope with an object glass of 28 inches diameter, form striking features in the landscape, and the public clock and standard measures of length in the outer wall of the Observatory are always objects of interest to visitors.

11.   People—Race.   Dialect.   Settlements.   Population.

London is the most modern of our counties, and yet as a British city it has behind it a history of nearly two thousand years. Alone among all our British cities, it has been of world importance through ten centuries, and at the present time it is the most cosmopolitan place in the whole world. In some of our English counties, such as Cornwall and Somerset, we find distinct traces of the speech and characteristics of the former inhabitants. This is also true of counties nearer London, such as Norfolk and Suffolk, where the Anglo-Danish influence is marked. But in London we have practically no definite survival of the original races which lived in the city; for the Londoner of to-day is either a recent immigrant from the country or from abroad, retaining his provincial or foreign characteristics, or else he is a hybrid of the most intricate ancestry.

It will not be necessary then to dwell at length on the various races that have lived in London. After its settlement by the Celts, there is no doubt that the Roman influence was of great importance, and that the natives were Romanised in many ways, as will be gathered from other chapters in this book. When the Romans withdrew, the English destroyed the city, and for a time it was a scene of desolation. The Roman villas, baths, bridges, roads, temples, and statuary were either destroyed or allowed to fall into decay. At length, however, London again became famous as the capital of the Old English kingdom of Essex, and continued to increase in size and importance. We may assign the renascence of London to Alfred, who repaired the buildings and rebuilt the walls. The building of St Paul's Minster in the tenth, and of Westminster Abbey in the eleventh century settled the question that London was to be a great ecclesiastical centre of the English people.

During the English settlement of London there were frequent incursions of the Danes, and they have left their mark on London, for several of the city churches retain their dedication to saints of Danish origin. Thus we have St Clement Danes church in the West, and St Magnus' and St Olave's churches in the East.

Perhaps the greatest change in London was effected in 1066, when the descendants of the Vikings, or

Northmen, who had settled in Normandy conquered our land. Then it was that William the Norman imposed his will on our people, and made London his capital. From the time that William built the White Tower, London has been without a rival, and has drawn to itself people from all parts of the British Isles and from all quarters of the globe.

Since the eleventh century no hostile settlement has been made in London, but there has been a steady influx of immigrants, who in many ways have added to the prosperity of the city. For we must remember that London has nearly always been kind to aliens, especially to refugees driven from their own land by political or religious motives. Flemings were brought over by some of our Anglo-Norman kings, and Germans from some of the Hanse towns became numerous in London in Plantagenet times. Then, too, in the sixteenth, seventeenth, and eighteenth centuries, the Protestant Huguenots were driven from France, and found a safe asylum in Spitalfields, Bethnal Green, and other parts, where some of their descendants still thrive. The Jews at various periods have made notable and valuable additions to the population of London, and have shown themselves supreme in the financial and other departments of the life of the metropolis. Italians have settled in considerable number in the Hatton Garden district, and the French are very numerous in the Soho neighbourhood.

During the latter part of the nineteenth century there was a steady immigration of Germans, Poles, and Russians, who settled mainly in the East End of London. These

Spitalfields Great Synagogue

(*Originally a Huguenot Church*)

aliens have not been an unmixed blessing, and an Act was recently passed to restrict the landing upon our shores of the poorest of these foreigners. The Act, however, is almost a dead letter, and the stream of undesirable aliens continues to flow, until now whole quarters in the borough of Stepney are practically inhabited by these people, who sell their labour at a very cheap rate. Their number is so large, that it is now necessary for policemen and other officials to learn Yiddish that they may deal with these settlers more effectively.

We may now pass to the question of dialect, and here our remarks as to the race of Londoners also apply. Owing to the influx of people from all parts of the British Islands, Londoners of to-day have no definite dialect as we should find in Yorkshire, or Cornwall, or Somerset. It has been well remarked that one of the most certain means of ascertaining the character of a people is afforded by their colloquial idioms. It would be very difficult to apply this principle to the speech of Londoners, which probably includes the worst as well as the best characteristics of our English language. Even in fashionable circles in London, we frequently find the conversation somewhat slangy, and given to the dropping of the final " g's " of certain words. Among the toilers of London, the speech is loud, and the initial " h " of words is seldom heard. The dialect of the Cockney has passed into a proverb, and has not only worked havoc in London, but has invaded the districts adjacent to the City, to the great detriment of the English language. Here it may be mentioned that the name Cockney is strictly applied to people born within the sound of Bow Bells: and perhaps no novelist has used the Cockney speech so largely as Dickens. In his early days it was common for Londoners to convert the letter "w" into "v," and *vice versa*, but this peculiarity does not now exist.

With these few remarks on the race and dialect of Londoners, let us turn our attention to the population of London as it was in 1911, when the last census was taken. For the statistics relating to the population we have no exact information till 1801, when the first census was made. From that date onwards there has been a numbering of the people every ten years.

In 1801 the population of London was 959,310 and in 1911 it was 4,522,961. This means that the population has increased nearly five-fold in the century. During the last twenty years the increase has been nearly 300,000, but the census of 1911 shows a decrease of 13,306. This enormous population is greater than that of either Scotland or Ireland, and exceeds that of several European countries. It forms about one-eighth of the entire population of England and Wales; and while the average number of people to a square mile in England and Wales is 618, it is no less than 38,690 in London.

The population of London north of the Thames is 2,678,651, and that south of it 1,844,310. There are 2,395,804 females and 2,127,157 males in London's population of 1911. The people lived in 1,019,546 separate tenements in 1901, of which the greater number, or 672,030, had less than five rooms. It was found that there were 40,762 single-room tenements, each having more than two persons, and 1602 single-room tenements with more than six persons in each.

The census figures are interesting in many ways. Thus we find that in 1901 there were 46,646 paupers in London's workhouses, and 4167 prisoners in the gaols. The military barracks of the metropolis held 10,058 people, and there were 10,675 inmates in the various hospitals. To give some idea of the longevity of Londoners, it is interesting to record that 161 people were between the ages of 95 and 100, while 24 people had exceeded 100 years.

The blind people of London numbered 3556, and these were largely employed in making articles of willow and cane, as brush-makers, and as musicians. The deaf and dumb were 2057 in number, and they worked as tailors, boot and shoe makers, and dressmakers.

Now we come to the census tables which give the place of birth of the people. We learn that 3,016,580 were born within the county of London, 35,421 in Wales, 56,605 in Scotland, 60,211 in Ireland, and 33,350 in British Colonies. Persons of foreign birth numbered 161,222, and were mainly natives of Russia, Germany, France, Italy, and the United States. The London borough having the largest foreign population is Stepney, where no less than 54,310 aliens were residing in 1901. The boroughs of Westminster, St Pancras, Holborn, and Marylebone have also a large population of foreigners.

As this book deals with the eastern portion of the county of London, we may close this chapter with a few figures giving some comparisons with regard to the populations of the various boroughs. Of the total population of London, 2,457,533 are in the eastern portion, against 2,065,428 in the western portion. Of all the London boroughs, Islington in the east has the largest population, viz. 327,423 people, while the City of London has the smallest, viz. 19,657 people. Of course the latter is the night population ; in the day-time the city population is ten times as great. Some of the boroughs in the eastern portion are densely populated, and Southwark, Shoreditch, and Bethnal Green have each 170 or more people to the acre. Woolwich and Lewisham have the least crowded populations, for the former borough has only 14, and the latter only 22 to the acre.

12. Industries and Manufactures.

London has long been celebrated for its manufactures as well as for its commerce. So early as the reign of Henry I the English goldsmiths had become so eminent for working the precious metals that they were frequently employed by foreign princes. The manufacturers of London in that reign were so numerous as to be formed into fraternities, or gilds, and of them we shall read something in another chapter. In 1556, a manufactory for the finer sorts of glass was established at Crutched Friars ; and flint-glass, equal to that of Venice, was made at the same time at the Savoy. Coaches were introduced in 1564, and in less than 20 years they became an extensive article of manufacture in London.

The making of "earthen furnaces, earthen fire-pots, and earthen ovens transportable" began in Elizabeth's reign, when an Englishman named Dyer brought the art from Spain. The same man was sent at the expense of the City of London to Persia, and he brought home the art of dyeing and weaving carpets. In 1577, pocket watches were introduced into England from Nuremberg, and almost immediately the manufacture of them was begun in London. In the reign of Charles I, saltpetre was made in such quantities in London as not only to supply the whole of England, but the greater part of Europe.

The Huguenots and other refugees from Europe brought their skill and instructed the people among whom they settled how to manufacture many articles that had been previously imported. Wandsworth became a busy little manufacturing town in 1573, when a colony of Huguenots introduced the hat manufactory; and it is said that Wandsworth was the only place where the Cardinals of Rome could obtain their hats.

The silk manufacture was first established at Spitalfields by the expelled French Protestants, after the revocation of the Edict of Nantes in 1685. For a long time silk-weaving was a most flourishing manufacture, and although it has much declined, there are still descendants of the old French Huguenots who live in Spitalfields and Bethnal Green. Foreign names are visible on the shop fronts, and some of the weavers still work in glazed attics such as their forefathers used in France. It is related that

Silk-weavers' Houses in Church Street, Spitalfields

(*Showing wide attic windows*)

the Pope in 1870 wished to procure a silk vestment woven all in one piece. Search was made in France and Italy for a man who could do this, but without success. At last he was found in Spitalfields, and, curiously, the weaver thus discovered was a direct descendant of one of the Huguenot refugees who had left France two hundred years before.

In the seventeenth century, Lambeth was a manufacturing centre, and foreign workmen taught our English people the art of making plate glass, delft ware, and earthenware. This last manufacture is still one of its principal industries, and Doulton ware is celebrated all over our country. It may also be mentioned here that Bow and Chelsea were justly celebrated at one time for their china, which is now eagerly sought for by collectors.

One of the most important industries in London is brewing, and it is interesting to know that the recipe for brewing was brought to London by some Dutch from Holland and Flanders. London stout and London ale have long been famous, and Stow tells us that in his time there were 26 breweries in the city all "near to the friendly water of the Thames." Nowadays there are upwards of 100 breweries in London, and many of them in order to get a good supply of water have sunk wells hundreds of feet deep. Some of the largest London breweries are in Southwark, Whitechapel, and Holborn. Besides brewing, distilling and sugar-refining are carried on in various parts of London, but the latter industry is rapidly declining.

Tanning and the leather trade have been carried on in Bermondsey and Southwark for hundreds of years, and are still in a flourishing condition. When these trades were first introduced, north-east Surrey had oak woods, but these have long passed away, and the whole of that district is now one of the busiest parts of London. Ber-

Cambridge University Press Warehouse

mondsey and Southwark have numerous other industries, and soap, candles, and biscuits are largely manufactured.

London has large manufactures of boots and shoes, and ready-made clothing, especially in the East End. A great deal of this work is carried on in the houses of the poor, and gives employment to hundreds of women and children.   The making of lucifer matches is also in the same district, and for such work the people are badly paid.

Industries connected with the printing trade, book-binding, and newspapers are of the greatest importance,

Printing-machine in *The Times* Office

and London stands at the head of all our towns in this respect.   The book trade at one time centred round Paternoster Row, but of late years many publishers have removed to the West End.   Most of the newspapers and magazines have their offices in Fleet Street and in that neighbourhood.

London is also one of the great centres of the furniture

and cabinet-making trades, and until recent times the chief shops and factories were in the neighbourhood of Shoreditch. These industries are declining, owing to foreign competition, and much of the work has been transferred to the neighbourhood of Tottenham Court Road and Oxford Street.

Royal Arsenal, Woolwich

Clerkenwell is famous for its clock and watch making, and Hatton Garden is the seat of the jewellery trade and the home of dealers in precious stones. Long Acre has a good deal of coach and carriage building; and at Lambeth there are important engineering works.

Although London is a large port, it has not much shipbuilding, though the Blackwall Yard can claim an antiquity of three centuries. Much that is now done is confined to the Isle of Dogs, a poor district which has also extensive docks and chemical works. London has always been celebrated for the manufacture of scientific instruments, such as those connected with surgery, optics, and mathematics, but now foreign scientific instruments are competing with London articles, and this trade is declining.

Woolwich is a district that owes its importance mainly to the Arsenal, which gives employment to as many as 14,000 workmen at a busy time. The artisans are employed in making all kinds of cannon, gun-carriages, shot and shell, rockets, fuses, and torpedoes. A great part of Woolwich is thus a crowded hive of workers, using their skill in one of the largest and most complete arsenals in the world.

13. Trade and the Gilds. The City Companies. The Markets.

Before locomotion by steam power, the geographical position of London made it the principal port of the whole island; and the introduction of railways and steamships in the nineteenth century has tended still further to increase its trade. From the earliest period of its history, London has depended on its trade for its supremacy, and Tacitus speaks of the concourse of foreign merchants in the London of his time. Although he does not tell us the nature of the trade, we know that there was corn to be shipped from the Thames, as well as tin, and oysters; while among the exports of the Roman occupation we may, without any doubt, reckon slaves.

When London was settled by the East Saxons, trade came to it again, and under Alfred and his successor it became the chief port and market-place of our land. Bede, who died in the eighth century, praises the happy situation of London on the Thames, and calls it the emporium of many nations. Among the names of London in the seventh and eighth centuries, we find *Ceap-stow*, *Lunden-Wic*, *Lunden-byrig*, and *Lunden tune's hythe*, and these certainly show the recognised importance of London as a market and port.

From the beginning of the ninth century the trade of London is more and more often mentioned. Then it is that the sea-faring merchant is rewarded, and the customs are of such importance as to be worthy of special regulations. It is probable that the first port was at Dowgate on the Walbrook, but as larger ships came, they moored alongside the wharves at Billingsgate and Queenhithe.

A blow was struck at the slave-trade in 1008, when it was decreed that "Christian men, and uncondemned, be not sold out of the country," and within a few years such merchandise was not necessary for the growing prosperity of the port. One remarkable fact in the early commerce and trade of London is that they were mainly in the hands of foreigners. Indeed, it has been said that London in those early times was largely a city of foreigners. In order to encourage our own people to trade, Athelstan, early in the tenth century, ordained that any merchant who had made three voyages should be acknowledged a thane.

Among the foreign traders who settled in London were some Germans, who were known to the English as Easterlings. They had their own hall, or Gildhall, which was called the Steelyard, and stood on the site of the present Cannon Street Station. These Teutonic, or Hanse merchants, had a monopoly of all the trade with the nearer countries of Europe, and they flourished in London till the reign of Elizabeth, when they lost all their special privileges.

The Italian money-lenders, known as Lombards, began to settle in London in the thirteenth century. They were mostly wealthy Italians driven from their own country, and, owing to the expulsion of the Jews from London, they did a large and profitable business. These Italian money-lenders have left their mark on London trade, and Lombard Street, named after them, is still the seat of our chief banking houses.

The reign of Edward III saw the increase of our trade with the Low Countries, and the settlement of Flemings in London and elsewhere. Flanders became the great market for our wool and so continued down to the time of the Tudors. Elizabeth gave a great stimulus to the trade of London, and it was in her reign that Gresham's Royal Exchange was built. When Antwerp was sacked by the Spaniards in 1576, London took its place as the leading port of Europe. The Hanse merchants lost their privileges by the action of Elizabeth, and at once the Merchant Adventurers took their place and carried on the wool trade with Flanders. It was owing to Elizabeth's wise commercial policy that other companies were formed for the development of trade. The Russian Company brought the furs of Russia to London, besides silks and teas from the East. The Levant Company developed a trade in the Mediterranean, and the East India Company began its work during the last year of Elizabeth's reign. All these trading companies enjoyed monopolies, but they brought increasing prosperity to the city, and in the next chapter we shall consider some of the measures which are now being adopted to enable London to maintain its premier position.

We can best grasp the present trade of London by looking at a few figures. The whole of the imports of the United Kingdom in 1907 were worth upwards of £645,000,000 and of this London took £210,000,000 or nearly one-third. The total exports of the United Kingdom for the same year were over £426,000,000, and London's share of this was about £75,000,000. It will thus be seen that more than a quarter of the trade of our country is carried on through London; and it will also be noticed that the imports exceed the exports by nearly three to one. London is therefore the chief port for imports, and is exceeded in its exports only by Liverpool.

Now it will be interesting to consider the chief articles that London imports. Nearly all the wool that comes to our country enters the Port of London, and its wool market is attended by buyers from all parts of the world. Most of the tea and coffee consumed in England is brought to London, and the city has practically a monopoly of the fur trade with Canada. London has a large share of the West Indian trade, which includes cocoa and sugar; and about a quarter of the tobacco trade belongs to it. Petroleum from America and Russia, cheese from Canada, and timber from the Baltic ports are almost monopolies of London. It is calculated that half of the wine that comes to England pays duty at the Custom House; and all kinds of French manufactures are sent to London. Among the chief exports of London are cotton goods, metal manufactures, wearing apparel, woollen goods, and machinery.

In dealing with the trade of London it is necessary to consider the Gilds and the City Companies. The gilds were meant to control the conditions of industry and to ensure reasonable rates. In the Middle Ages few men and women could stand alone, and combination was a necessity for existence. One object of the gilds was the maintenance of good quality in the goods produced, and the gilds were responsible for their members. There were craft gilds and merchant gilds, but it is difficult to ascertain the relation of the former to the latter. Some of the merchant gilds appear to have been specialised into gilds of particular crafts, and craftsmen were sometimes members of them. There were three classes of gild members: the apprentices, whose relations with their masters were most carefully regulated; journeymen and servants; and masters.

A large number of the early gilds were purely social, and there is no trace of merchant gilds before the Conquest, while craft gilds did not come into existence until early in the twelfth century. In the year 1180, there were eighteen gilds in existence, and among them were the Goldsmiths, the Pepperers, and the Butchers. Gradually the influence of the craftsmen made itself felt, for they found a patron in the Mayor of London. There are records of the regulations of some of the gilds in the fourteenth century, and from these we find that the members had to be of good repute, and take an oath on entry, besides giving a kiss of love, charity, and peace. The members promised to nourish good fellowship, and men of evil life were put out of the fraternity. We also find that members received weekly help in poverty, old age, and sickness, while the young were helped to get work. The whole story of the City Gilds is of deep interest, but here we can only mention further that they were abolished in 1547. A writer on the subject says that "no more gross case of wanton plunder is to be found in the history of all Europe; no page so black in English history."

Now when we pass to the present Livery Companies of the City of London it is not easy to decide whether they were in any way related to the craft gilds. There seems, however, no doubt that the old craft gilds and the later companies were very closely connected, for they were both formed to promote the combination of the sections of a particular trade. In addition to this the City Companies controlled municipal politics, and even to this day their members, or Liverymen as they are called, elect the Lord Mayor of the City of London. Their supremacy over the different trades has, however, nearly all gone, although the Fishmongers, the Goldsmiths, the Stationers, and the Apothecaries have still some authority in those particular trades.

Goldsmiths' Hall: the Grand Staircase

The Livery Companies, with their political and municipal power, are peculiar to London, for no other city has permitted associations of this kind to amass such wealth and power. There are now seventy-six City Companies, and of them twelve are known as the Greater Companies, whose yearly incomes range from £11,000 to £83,000. The names of the more important companies are the Mercers, Grocers, Drapers, Fishmongers, Goldsmiths, Skinners, Merchant Taylors, Haberdashers, Salters, Ironmongers, Vintners, and Clothworkers. The richest of them all is the Mercers, to whom belonged Whittington, Sir Thomas Gresham, and other famous City merchants; while the Merchant Taylors contains the name of every English king from its foundation. These City Companies spend their money on the many charitable institutions they maintain in London; they contribute most generously to various benevolent objects; and they maintain and support many schools, besides showing great interest in promoting technical instruction in all parts of the County of London and in the older Universities.

We will now pass from the regulation of trade by the Gilds and City Companies to a consideration of the Markets of London. Markets have been in existence in the City for more than a thousand years, and the City Corporation has for many centuries been the market authority for London. The City in the reign of Edward III was granted exclusive market rights and privileges within a radius of seven miles, and these rights have from time to time been recognised and confirmed.

The Central Markets at Smithfield cover part of the

site of Old Smithfield Market, which was founded for the sale of live stock. The present markets for the sale of meat, poultry, and provisions are strictly wholesale, except on Saturday afternoons, when the "People's Market" is held, and a large business with the poorer classes is carried on. It is calculated that about 430,000 tons of meat are sold here in one year.

Smithfield Meat Market

The Metropolitan Cattle Market at Islington deals with the sale of oxen, sheep, pigs, and calves bred in our own country. Here, in 1907, no less than 455,000 head of live stock were sold. The market is held twice a week, but the lairs are open for the reception of animals at all hours of the day and night.

The Foreign Cattle Market is at Deptford, and stands on the site of the old Dockyard. It has an area of upwards of 30 acres, and accommodation for 8,000 cattle and 20,000 sheep. Every animal is inspected on its entry by veterinary officers, and slaughtered within a period of ten days. The diseased animals are at once consigned to a digester, and reduced to ashes.

Billingsgate is London's chief fish market, and dates from time immemorial. The market is for the wholesale and retail sale of fish, which arrives both by land and water. The water-borne fish, caught chiefly in the North Sea, is collected from the various fishing-fleets by steam vessels known as steam carriers, which deliver at Billingsgate Quay. In 1907, as much as 174,000 tons of fish were sold by auction at this noted market.

Leadenhall Market has existed from very early times, and sells meat, poultry, game, and provisions. The present market was opened in 1881, and the business is both wholesale and retail. At Leadenhall Market all kinds of pets may be bought, and a large trade is carried on in such animals as goldfish, hedgehogs, dogs, foxes, parrots, tortoises, and rabbits.

The Smithfield Hay Market has practically fallen into disuse, although an open space is still preserved at Smithfield for the sale of hay and straw. Spitalfields Market is largely for the sale of vegetables, and Shadwell Market for the sale of fish.

Besides the above markets, which are managed by the City Corporation, there are Whitechapel Hay Market under the control of Stepney Borough Council; the

Borough Market partly managed by the Southwark Borough Council; and the Woolwich Market under the control of the Woolwich Borough Council.

14. The Customs and the Custom House. The Exchanges. The Bank of England. The Royal Mint.

Customs' duties are that portion of the national revenue which is derived from a tax on imports, and as London is the chief port for imports, more than half of the customs' duties are collected at the Custom House. The tax on imports was of old a simple percentage, which was known as "tunnage and poundage," from the method in which it was levied on the *tun* of wine, or the *pound* of other merchandise. These sums were granted first to the Crown, and then for the maintenance of the State.

Many changes have been made from time to time in levying these duties, and before the Free Trade legislation of 1846 it was computed that taxes were collected on 443 articles. At present our tariff contains only twenty dutiable imports, and the whole customs' revenue is practically derived from spirits, wine, tobacco, coffee, tea, and dried fruits. The collection and general management of British customs' duties is under one great central government department in London, and the Custom House in Lower Thames Street is the building where the customs are collected for the Port of London. The present building is the fifth on this site and was erected in 1814. It has a river frontage of 488 feet, and the quay forms a noble esplanade with a fine view up the river. The Long Room, in the centre of the building, is considered to be the finest and longest of its kind in Europe : and besides extensive warehouses and cellars,

The Custom House

there are upwards of 170 apartments in the building. Goods that are seized or forfeited are stored in the King's Warehouse, and when it is full there is a public sale of the goods.

Passing from the Custom House, we may proceed to consider some of the Exchanges of London, and we cannot do better than begin with the Stock Exchange. It is situated in Capel Court, between Threadneedle Street and Throgmorton Street, and is quite close to the Bank of England. It is known in the City as "the House," and numbers some 3000 members, who have the right to buy and sell stocks. Visitors are not admitted, and if a stranger happens to find his way in, he is roughly handled and hustled out. London has been called the

The Royal Exchange

"bank of the whole world," and certainly the Stock Exchange has business of a cosmopolitan character, for its members deal in all kinds of home and foreign stocks and shares.

Not far from the Stock Exchange is the Royal Exchange, one of the best known buildings in the City. It is the third exchange built on this site, and is in the

classic style, with a portico of Corinthian pillars, surmounted by a pediment, and based by a broad flight of
steps. The first Royal Exchange was built by Sir Thomas
Gresham, one of London's merchant princes in the reign
of Elizabeth, and his crest, the grasshopper, still adorns
the building as a vane on the top of the east tower.
Gresham got the idea of the first building from the
" Bourse " of Antwerp, and he wished English merchants
to have a similar house where they might meet and
transact their business. Gresham's Exchange was opened
by Queen Elizabeth in 1571, and the present Royal
Exchange was opened by Queen Victoria in 1844. In
the centre of the large enclosed court, which has the
tessellated pavement of the first Exchange, there is a
statue of Queen Victoria. The walls of the court are
now being decorated by paintings illustrating the chief
events in the history of English commerce. It may be
mentioned that the first Royal Exchange was destroyed
in the Great Fire of 1666, and the second, which was
opened in 1669, was also destroyed by fire in 1838.

The eastern part of the Royal Exchange is occupied by
"Lloyd's," which is short for Lloyd's Subscription Rooms.
" Lloyd's " was founded by Edward Lloyd in the seventeenth century, and is now an association of underwriters,
whose business has largely to do with shipping and
insurance. Lloyd's *Register* is a most important annual,
for it classifies our steam vessels and sailing vessels and
distinguishes them by letters and figures such as "A 1,"
this well-known symbol denoting the highest class of
vessels.

Besides the Stock Exchange and the Royal Exchange there are several other Exchanges in the City for business in special articles. Among them may be mentioned the Corn Exchange in Mark Lane, the Shipping Exchange in Billiter Street, the Wool Exchange in Coleman Street, and the Coal Exchange in Lower Thames Street.

We now come to the last section of this chapter, which will deal with the Banks, whose chief seat for a long time has been in Lombard Street. Here it was that the great medieval money-lenders from Italy settled hundreds of years ago, and here it is to-day that a large part of the trade and exchange transactions of England takes place. Of course there are many other banks in London besides those in Lombard Street, but there, and in the streets near it, are the headquarters of some of the greatest banking establishments in the world.

When we speak of the banks of London, our mind naturally turns to the Bank of England, which is often called the "Old Lady of Threadneedle Street," because of its position. The Bank of England is one-storeyed and irregular, and loses in effect from not being raised on a terrace. It has no exterior windows, but for the sake of security is lighted from within. The Bank of England was founded by William Paterson, a Scotsman, in 1694, and the first governor of the Bank was Sir John Houblon. The Bank transacts much business for the Government, which may be said to be its chief customer, and it is the only bank in London which has the power of issuing paper money. The Bank vaults contain about £20,000,000 in gold and silver, and for protection a military guard is every night stationed within the walls. The business of the Bank is open to the public, but to see the printing, weighing, and bullion offices, special orders are necessary. The whole of the printing of the Bank is done within its walls, and besides the bank-notes, postal orders are produced. In the weighing office there are machines for weighing sovereigns at the rate of 33 per minute,

The Bank of England

those of full weight being thrown into one compartment, while the light ones pass into another. The bullion office is the treasury for the precious metals, and the Bank is bound to buy all gold bullion brought to it at the rate of £3. 17s. 9d. per ounce.

The Royal Mint on Tower Hill is connected in some ways with the Bank of England, for it is here that the process of coining takes place. All the gold bullion from the Bank of England is here made into sovereigns and half-sovereigns, and in addition all our bronze and silver coins are struck here. The processes of coining are most interesting, and the machinery used is of a most ingenious character. There are some presses which can

The Royal Mint

stamp and mill 120 coins per minute, and one of the machines into which gold coins pass separates them into proper weight, light weight, and heavy weight. The Mint museum contains a large number of cases of coins and medals which have been in use in our land from the earliest times. The Royal Mint also makes coins for other countries, and besides gold, silver, and bronze, it

also uses nickel and aluminium.   The value of our
Imperial coinage issued by the Mint in one year amounts
approximately to £7,000,000.

15. The Port of London—The Docks. Shipping. Shipbuilding.

London has always stood first among British seaports
and is now the greatest seaport in the world.   The
reasons for this pre-eminence are mainly because of its
geographical advantages.   It is situated on both sides of
the Thames, which is a broad navigable river, running
far into the land, and having a tide which twice a day
carries vessels into its docks.   It stands opposite the
great markets and chief ports of the continent, and this
nearness to the continental ports is probably the main
cause of its great *entrepôt* trade.   The great value of its
imports and the extent of its *entrepôt* trade are two special
features of its foreign trade.   With regard to the value
of its exports, as has been seen, it is surpassed by Liverpool,
owing to the more favourable situation of the latter with
reference to the great manufacturing districts of England.
The vast import trade of London is largely due to its
enormous population, which consumes so much of the
food brought up the Thames.

For some years past, however, there has been a feeling
that London as a port had not been making the same rate
of progress as some of the continental ports, where docks

have been built which rival those of London. Hamburg
and Antwerp have grown with remarkable rapidity, and
Bremen, Rotterdam, and Havre are all competing with
London and other British ports. In these foreign ports
everything has been done to attract shipping, and large
sums of money have been spent to make their docks
accessible, convenient, and cheap. The best machinery
for loading and unloading ships has been provided; large
and convenient wharves have been built; and cheap
fares on the railways communicating with the docks are
the rule.

Various proposals were made from time to time to
improve the Port of London, and in 1903 a Bill was
introduced for putting it under public control. This
measure, however, was not carried, and it was not till
1908 that another Bill was introduced " to provide for
the improvement and better administration of the Port of
London and for purposes incidental thereto." This Bill
was favourably received and in due course became law.

Among the provisions of this measure was the establishment of the Port Authority, consisting of 18 elected
and 12 appointed members, who were to administer,
preserve, and improve the Port of London. The Port
Authority has taken over the docks, and has power to
construct, maintain, and manage any docks, quays, wharves,
jetties, and piers that may be necessary.

Before 1908, the Thames Conservancy had control
over the Thames from its source to the sea, but that
body is now restricted in its work to the eastward
limit of Teddington Lock. The Port Authority has now

jurisdiction over the Port of London, which extends from Teddington Lock, 19 miles above London Bridge, to an imaginary straight line drawn from Havengore Creek in Essex to the Land's End at Warden Point in Sheppey, Kent.

Besides the new Port Authority, there are two other bodies which have jurisdiction in particular cases over the same area. First there is the City Corporation, which is the Port Sanitary Authority from Teddington Lock seawards; and secondly there is Trinity House, which has charge of the pilotage, lighting, and buoying of the river from London Bridge seawards. Pilotage is compulsory in the London district for vessels exceeding 60 tons burden, engaged in foreign trade. The fees for pilotage are paid according to the distance piloted and the draught of the vessel, and Trinity House receives 6d. in the £ on the earnings of the pilots, and an annual fee of three guineas on the renewal of each pilot's licence. There are about 350 licensed pilots in London and the amount received for pilotage in one year amounts to nearly £150,000.

The Thames requires considerable skill in navigating, for the navigable channel at the Nore is only 1000 feet in width and it decreases to 200 feet at London Bridge. The depth of the navigable portion varies from 26 feet to as little as 14 feet. It will therefore be seen that the work of lighting and buoying the river is of the greatest importance, and Trinity House has a large number of lights, buoys, fog-horns, bells, and fog-sirens to warn mariners of the sandbanks. At the entrance to the

East India Docks

estuary is the Nore light-vessel, which is in four fathoms of water at the east end of the sand. It shows a white light, which revolves every half minute. The vessel has a red hull with the name *Nore* painted on both sides and with a ball at the masthead.

Now leaving the navigation of the Thames, let us briefly consider the Docks, which were taken over in 1909 by the Port Authority. On the north side they are as follows :—St Katherine's, London, West India, Millwall, East India, Royal Victoria, Royal Albert, and Tilbury; and on the south side are the Surrey Commercial Docks. Altogether these docks have a water area of 645 acres, and they are all closed docks, being entered by locks.

The West India Dock was the first dock in London and was opened in 1802. Before that date, ships were unloaded at certain quays and wharves, but the accommodation was quite insufficient. After the first docks were opened, their value was realised, and others followed in quick succession. The great deep-water docks at Tilbury were the last, and were opened in 1886. The largest ships in the world can enter them at any state of the tide, and each of the two graving docks has a length of 846 feet.

Besides these docks from London Bridge to Tilbury, there are, along this water-way of 25 miles, a number of landing-places and wharves where ships can discharge their cargoes. Very large vessels come up-river to London Bridge, but the great ocean liners do not come so far, for they generally berth in some of the large docks lower down.

Royal Victoria Docks

In order to give some idea of the vast trade of the Port of London we may mention that, in 1907, 25,857 vessels entered, and 26,474 vessels cleared. The tonnage of the former amounted to over 17 million tons, and of the latter to about 16½ million tons. The value and character of the goods has been already stated, and we may close this portion of our chapter by remarking that there is every reason to believe that under the new Port Authority the trade will increase with the additional advantages offered to shipping.

Although London is such a great port, it is by no means important as a shipbuilding port. Indeed, it ranks very low, and year by year this industry is declining; and some of the large shipbuilding firms have removed their works from the Thames to the north of England and to Scotland. At present there is some shipbuilding at Poplar, and at the Thames Iron and Shipbuilding yards at Canning Town. In 1906, out of 1500 vessels built in the United Kingdom only 150 were built in London. The tonnage of the latter was but 10,507 against 1,148,500 of the former.

16. History.

There is no authentic evidence to show at what period the Britons settled in the district we now call London. We have already referred to the origin of the name, and also to the probable position of the city in the days of the

Britons. Here we need only remark that it was then probably little more than a collection of huts on a dry spot in the midst of a marsh, surrounded by an earth-work and ditch. It is not possible to say what length of time elapsed between the foundation of British London and its settlement by the Romans. We do know, how-ever, that after the Roman conquest London rapidly grew in importance, and in A.D. 61 it is spoken of as a place noted for its concourse of merchants.

The early Roman city on this site was called Augusta, and was founded in the reign of Nero, A.D. 62. As a Roman city it did not rank in importance with either Eboracum (York) or Verulamium (St Albans), and it was never regarded as the capital of Roman Britain. Under the Romans the city extended from the site of the present Tower of London on the east to Newgate on the west, and inland from the marshy banks of the Thames to some swampy land known as Moorfields. The Romans left their mark on London, and the wall with its gates and the bridge over the Thames which they built are all referred to in other chapters.

The actual historical references to London during the Roman period are very few. In the year A.D. 61 Boadicea, the British queen, attacked the town, and Suetonius and his troops were forced to evacuate it. For more than two centuries after its capture by the British forces, we have no mention of London by any historian. There is every reason to believe that the Romans recaptured it, and that its prosperity increased. Towards the end of the third century, Carausius, the

Roman commander in Britain, proclaimed himself emperor, and struck in London gold coins bearing his portrait
and name.   Before long, however, he was murdered by
Allectus, who assumed the imperial title, but was defeated
and slain at Southwark by a general whom the Emperor
of Rome had sent against him.   Henceforward the history
of London becomes fragmentary.   In 410 the Roman
soldiers were withdrawn from Britain, and with their
departure the Roman name of the city, Augusta, was
forgotten or, at all events, disappeared.   When the city
comes again into the light of history, it is under its more
ancient name—London.

This silence of history for two centuries is very remarkable, for we hear of the English conquest of Pevensey,
Bath, and Gloucester, but the story of London is quite
lost to us.   In the year 604 A.D. we find the city in
the possession of the East Saxons, and new names are
given to the old Roman roads, the gates, the rivers,
and the hills.   Everything is changed, and the power
of Rome over London has vanished.   London became
the capital of the kingdom of the East Saxons, and
continued to increase in size and importance.   As early
as the beginning of the seventh century the influence
of Christianity made itself felt in London.   Ethelbert,
king of Kent, had been converted, and as overlord of all
nations south of the Humber he had a sincere desire that
the East Saxons should become Christians like the people
of Kent.   He therefore decreed that the people of London
should put away the worship of Thor, Odin, and other
gods of the north.   Evidently he was obeyed, and Mellitus

was consecrated the first bishop of London in 604. Bede
tells us that Ethelbert built the first church of St Paul in
London, and the site of the present Westminster Abbey
was also occupied by a Christian church. Many other
churches were built in various parts of London, and were
dedicated to national and local saints, such as St Dunstan,
St Botolph, St Osyth, and St Swithin.

From the time of its conversion, London steadily
grew, and learning and culture came from over the sea
to its people. Monasteries were founded, and monks
from the continent came to fill them. The arts of
architecture, painting, and music developed, and a brighter
and better life for the inhabitants of London was the
result. The Christians of the seventh and eighth centuries were quite unlike the pirates and plunderers of
earlier days. Peace had settled on the land, and, as a
result, commerce brought riches to London. Early in
the ninth century, however, a new enemy appeared at the
gates of London, and for a long time the Danes harried
the city. At length, in 839, they captured it, and made
it their headquarters. The task of the Danes was com-
paratively easy, for as the English were not distinguished
as builders, they had not strengthened the fortifications of
London.

At the time of the occupation of London by the Danes,
Alfred was king, and he made it his aim to recapture the
city. He recognised the value of London as a possession,
and in 884 the city fell into his hands. The name of
Alfred is imperishably connected with London, and one
of our historians goes so far as to say that Alfred gave us

Staple Inn, Holborn

London. At any rate, in the year 886, Alfred determined to rebuild and strengthen the city. The *English Chronicle* says, "Alfred honourably rebuilt the city of London and made it again habitable." Never again did the Danes conquer London, and from that date London has continued to increase in wealth and prosperity.

The Danish occupation of London may be traced in some of the names of its churches and streets. The churches dedicated to St Magnus, St Olave, and St Clement Danes remind us of the Danish settlement. Tooley Street is a corruption of St Olaf's Street, and Gutter Lane, off Cheapside, is said to be the modern form of Guthrum's Lane.

The last event of importance in early London was the building of Westminster Abbey by Edward the Confessor. This was not the Abbey that we see to-day, but a church built on what was then the swampy island of Thorney, and around it grew up the city of Westminster.

During this period of our history, London fought an uphill fight with Winchester for the position of chief city of England, and it was not till the reign of Edward the Confessor that it became the recognised capital of our country. This position was still further secured when William was crowned in Westminster Abbey on Christmas Day, 1066, and in return for the submission of the citizens, William granted them a charter, which was of immense importance to them. Having gained possession of London, William proceeded to fortify the city. He enclosed a space of about twelve acres in the east of London, and gave orders to build the Tower. Gundulf,

Bishop of Rochester, was the architect, and the White Tower as we see it to-day is the most important remnant of Norman London.

In the year 1100, Henry I granted the citizens of London a second charter, and from the advantages it conferred we may measure the growing importance of the city. We have the first record of the Mayor of London in 1190, when Henry Fitzailwin was elected to that high position, which was held by him for twenty-four years. London was recognised as a *communa*, or fully organised corporation in 1191, and when King John was quarrelling with the barons, the Londoners took the side of the latter. The Magna Carta specially secured to London its rights and customs as obtained in previous charters, and gave it the power to elect its mayor annually. It also ordained that the Mayor of London should aid the twenty-four lay peers to compel the king, by force of arms, to keep the Charter.

The reign of Edward III was remarkable in the history of London, for royal charters were granted to some of the craft-gilds, and henceforward the Livery Companies had a direct share in the government of the city. The power of these companies was enormous, and it was mainly owing to them that London became the first industrial and commercial city in the kingdom.

The Londoners of the time of Richard II showed their power, for they refused the king a loan, and though he deprived them of their charters, it was not long before they were restored. When Wat the Tiler and his followers entered London in 1381, John of Gaunt's palace in the

Savoy was burnt, and some of the prisons were opened. While Richard himself was meeting some of the insurgents at Mile End, a large body of them broke into the Tower and murdered the Chancellor and the Treasurer, and other officials. Next day, Richard ventured again to meet them. Wat the Tiler, their spokesman, was so insolent that Walworth, the Mayor of London, cut him down. The angry multitude were dispersed after the king had promised that their grievances should be remedied.

King Henry V entered London in triumph in 1415, after his great victory at Agincourt. Richard Whittington, the mayor, entertained him at a banquet at his own private house, and the citizens were most enthusiastic in their reception of the victorious monarch. The king attended a thanksgiving service at St Paul's, and then retired to his palace at Westminster.

The reign of Henry VI is memorable in the annals of London for the capture of the city by Jack Cade in 1450. With a large force of Kentishmen, Cade entered London without meeting with any resistance, and, riding up to London Stone, he struck it with his sword, exclaiming, "Now is Mortimer lord of this city." For three days his followers plundered and burnt, until by the exertions of the mayor and aldermen, the rebels, who had retired to Southwark, were shut out of the city.

When the Wars of the Roses began, the Londoners espoused the cause of the Yorkists, and when Edward of York appeared before the gates of London, the citizens received him with acclamation. A little later, the people met one Sunday in an open space near Clerkenwell,

and with the familiar shout of "Yea! Yea!" chose
Edward IV to be their king. To the day of his death
this monarch was popular with the citizens of London.
The story of London under the Plantagenets ends with
the reign of Richard III, who like his brother, Edward IV,
was invited by the mayor and chief citizens to be King
of England.

During the Tudor period London continued to grow
in importance. Reference will be made in another chapter
to the dissolution of the religious houses in the reign of
Henry VIII, but here we may note that London was the
chief scene of the burning of "heretics" at Smithfield in
the reign of Mary. When Elizabeth was on the throne,
the capital showed its patriotism by its liberal contributions
of men, money, and ships for the purpose of resisting the
threatened attack of the Armada.

Under the Stuarts the history of London assumes
even greater importance. Owing to the exactions of
the Star Chamber it sided with the Roundheads, and
became the centre of Presbyterianism and of opposition
to the king. In 1648, the city was occupied by the
Cromwellian troops, and next year Charles I was beheaded at Whitehall. Cromwell was proclaimed Lord
Protector of England in 1653, and after his death London
was occupied by Monk's troops. The year 1660 witnessed
the Restoration of Charles II, who was received back to
his kingdom with the greatest satisfaction by Londoners.

The reign of Charles II is memorable in the history
of London for two great events. The Plague of 1665
turned the capital into a city of mourning and desolation,

The Monument

and it is calculated that about 100,000 Londoners died of
that fell disease. The following year witnessed another
dire calamity, for the Great Fire destroyed no less than
13,000 houses and 89 churches. This disaster, however,
proved beneficial to London, for the city was rebuilt in an
improved form. Its streets were widened, and the wooden
houses gave place to buildings of stone and brick. The
Monument on Fish Street Hill was finished in 1677 as a
memorial of the Great Fire. It is 202 feet high, and
stands at a distance of 202 feet from the spot where the
fire first broke out on September 2, 1666.

It was not till the reign of Queen Anne that London
began to assume anything like its present appearance. It
was during her reign that the results were evident of
Wren's rebuilding of the Cathedral of St Paul's, and
many churches in the city. During the eighteenth
century London increased in size and population, and
during the latter part of that period some of its hand-
somest streets were made, and some of its finest buildings
erected. Great injury was inflicted on the city by the
Gordon Riots of 1780. Lord George Gordon put
himself at the head of Londoners with the cry of
" No Popery ! " Prisons were destroyed, the prisoners
released, and mansions burned or pillaged. The rioters
were not subdued till hundreds of them had been killed
and Lord George Gordon sent to the Tower.

An important event in the social life of the city was
the lighting of the streets with gas. Pall Mall was
the first street thus lighted in 1809, and Bishopsgate
Street followed in a short time. The story of London

throughout the nineteenth century is one of remarkable growth and expansion, and London has now made good its claim to be not only the largest, but also one of the finest cities in the world. The Metropolitan Board of Works was formed in 1855 to look after the sanitary arrangements of London, and in 1889 this body gave place to the London County Council, whose aim must be, in the words of Lord Rosebery, " to make it more and more worthy of its central position, of its great history, and of its immeasurable destinies."

17. Antiquities—Prehistoric, Roman, Saxon.

The conditions of man's existence on the earth during the early stages of his history are shrouded in obscurity. The earliest evidence of the presence of man in London is not by written records, but by implements of chipped flint. The very earliest of these implements belong to a time when our country was joined to the continent, and their age must be reckoned by thousands, if not by tens, and possibly hundreds, of thousands of years.

Antiquaries have divided this early period of our country's history into the Stone Age, the Bronze Age, and the Early Iron Age. It must not be thought that, in the Bronze Age, stone was discarded for many purposes, or that bronze was no longer in common use after the discovery of iron. On the contrary, each material survived long into the succeeding period; but

this classification is convenient as it shows the material
chiefly in use. These three Ages or Periods cover a vast
extent of time, but it is not necessary to say how many
years are included in each of them, for we cannot be
certain when one age ended and the next began.

The Stone Age has been further divided into the
Palaeolithic or older section, in which the flint implements were formed simply by chipping, and the Neolithic,
or newer section, in which they were more carefully
worked, and even polished. There is reason to suppose
that an immense period of time separated the two Ages.
Palaeolithic implements had no doubt been found before
it was recognised that they belonged to a remote past ;
but the first recorded discovery of the kind was made in
London towards the end of the seventeenth century. A
fine pear-shaped implement was found with an elephant's
tooth near Gray's Inn Road, and was described, wrongly
of course, as a British weapon.

As London lies in the valleys of the Thames and
Lea, it is not at all remarkable that so many traces
of man are found within the borders of the county.
Both river-valleys have yielded many examples of flint
weapons and implements of the Palaeolithic Age. In
the bed of the Thames a good many specimens have been
found at various times, and they are frequently brought
up in the course of dredging operations.

Palaeolithic sites have been traced at Stoke Newington,
Stamford Hill, Kingsland, and in the City of London, and
many specimens of flint implements from these localities
are to be seen in the British Museum. Some flakes found

at Stoke Newington Common and struck off by palaeolithic man from the same core are shown fitted together
again, as evidence that the manufacture of flint implements
took place on this site.

Palaeolithic Flint Implement found in Gray's Inn Road

In the Guildhall Museum there is a fine collection of
London antiquities, and among them are implements from
the river-drift gravels that have been discovered at Bishopsgate, Wandsworth, Clapton, and other places. The very

best way of learning about stone implements is to visit either the British Museum or the Guildhall Museum, and there, with a good guide, to examine these interesting relics of antiquity.

When the Neolithic Age began, Great Britain had ceased to be a part of the continent and was an island. The climate had become more temperate and rather moist, while such animals as the mammoth had become extinct. Man had now learnt to train animals for domestic use; and he cultivated cereals for food, and various plants to provide materials for woven garments. He used the bow as his weapon, and he had also developed the art of making pottery.

In the Neolithic Age implements and weapons were commonly hafted and made in a greater variety of forms; and by the addition of grinding and polishing, it was found possible to use other hard stones besides flint. While the grinding and polishing of stones may be considered the special characteristic of this period, it must not be supposed that this was always the case. For instance, a large and important class of implements and weapons, such as knives, scrapers, and arrow-heads, were but rarely ground or polished, while even axes of fine workmanship were sometimes finished by simply chipping them.

Among the implements found in London belonging to the Neolithic Age may be mentioned celts from Paddington and Southwark; scrapers from London Wall and Battersea; a flint knife from Addison Road, Kensington; and flint flakes from Fulham and Hammersmith.

As the Neolithic Age advanced, man gradually learnt the use of metal, and from this important step in human progress we find traces of his rapid advance in all directions. The period from the beginning of metallurgy down to the dawn of written history is generally divided into two parts :—an earlier or Bronze Age, and a later Age of Iron. Among the antiquities found in London belonging to the Bronze Age may be mentioned a bronze sword of leaf form from the Victoria Embankment, a bronze dagger-blade from the Thames, an implement of red deer-horn from Philpot Lane, and a fragment of pottery ornamented with herring-bone pattern from Hammersmith. Besides these antiquities, socketed celts, winged celts, and palstaves have been found in various parts of the county of London.

Under the portico of the British Museum are some "dug-out" boats which possibly date from the Bronze Age. They belong to a common type, formed out of a tree-trunk split lengthwise, the work of hollowing the interior being performed by tools of stone or bronze, and probably by fire. One of them was found during excavations for the Royal Albert Dock at North Woolwich in 1878. The oak trunk was carefully worked, the bottom and sides being flat and rectangular, but there are no signs of keel, stretchers, or rowlocks.

When the knowledge of iron and the valuable properties it possessed became known in Britain it is probable that the new metal supplanted bronze in the manufacture of such implements as sword-blades, daggers, and knives. Iron is believed to have been brought into our country by

the Brythons, a branch of the Celtic family, and from
them our island received its name of Britain. During the
Iron Age, man in Britain made great progress in culture
and the arts, and the antiquities of this period bring us

Enamelled Bronze Shield (*from the Thames at Battersea* (⅛))

down to the conquest of Britain by the Romans.   Many
articles of great interest belonging to the Early Iron Age
have been found in London, and they include all kinds of
personal ornaments, such as fibulae, hairpins, and rings,

various implements and weapons of iron, such as knives, swords, and daggers, and many specimens of pottery, such as vases and urns.

A bronze shield decorated with red enamel was found in the Thames near Battersea, and an examination of this impresses one with the beauty of the curves and the well-balanced proportions of the various parts. The boss of another shield was also found in the Thames near Wandsworth, and the decorations on it are produced by means of nearly complete circles, enclosing leaf-like thickenings. Both these examples represent the very best work of the Early Iron Age, but there have been found other articles of great merit. A bronze brooch from the Thames, and a scabbard with mock-spirals from the Thames at Wandsworth, are also extremely interesting specimens of this period.

Before we leave this early Iron Period, it may be well to ask what monuments are left in London to remind us of the early Celtic people in our land. Of prehistoric monuments in and around London there are only two, and they are the Hampstead Barrow and the River Walls. The Hampstead Barrow, locally called Boadicea's Grave, is of doubtful origin. It was opened in 1894, but nothing was found in it, and it may be only a boundary hillock. With regard to the river walls, which may be seen from Barking on to Southend, the tradition is that they were first built by the Britons. There is no date for them, and, although they have been often repaired, they are practically the same as when first constructed. We found in the first chapter that the word London may remind us of the Britons, the early Celtic people in our land. If the word London is derived from Caer-Lud after a King Lud of Celtic history, then we have in the present name of the county a direct link with this early period of our history.

The Roman remains found in the county of London are very numerous, and from them we are able to picture to ourselves the life of its people two thousand years ago. It has been said that the two chief events in the history of Roman London are the building of the bridge and the building of the walls, but as both of these are considered in other chapters we need make no further reference to them here.

Roman London may be termed a buried city, for all its remains have been found many feet below the level of the present streets. Among the smaller antiquities found have been personal ornaments, such as bronze fibulae, rings, brooches, hairpins, earrings, and gems, and many of these articles have been disinterred in various parts of the City of London, especially in London Wall and Aldgate. Domestic utensils and appliances are frequently unearthed, such things for example as Roman balances, bowls of bronze, forks, knives, and spoons, which have been found near the Mansion House, in Queen Victoria Street, and in London Wall. Iron lamps and lamps of glazed ware have been unearthed in Lothbury and Broad Street, and all kinds of tools, such as chisels, axes, adzes, and piercers have been discovered in London Wall and at Wapping. Very numerous discoveries have been made of metal objects such as bells, chains, and horse-furniture at Southwark, Austin Friars, and Walbrook.

Among the more interesting Roman remains in London may be mentioned figures and statuettes in metal, clay, and terra cotta. In the Guildhall Museum there is a very fine collection of these objects, and statuettes of Apollo, Hercules, Juno, Mercury, Mars, and Venus are quite numerous. From them we get a good idea of Roman art, and also a realisation of the chief Roman deities.

From the Roman remains in London we can picture to ourselves the internal decorations of the houses of the citizens. No less than 40 mosaic pavements, either complete or incomplete, have come to light, and of these the most beautiful and perfect is the mosaic pavement found in Bucklersbury in 1869. It is composed of red, white, grey, and black tesserae; the lower and main portion is in the form of a parallelogram, while the upper part is semi-circular. The central device is a floral design, surrounded by a cabled band, and enclosed within two squares of ornament placed at different angles, having a floral device at each corner. Between the upper and lower portions is a broad band of floral scrolls. The upper semi-circular portion has a fan-shaped device in the centre, above which is a pattern of scale ornament, and the border consists of a knotted band. The whole design of this fine mosaic pavement is enclosed in a border of red tesserae.

Roman glass and pottery have been found very extensively in various parts of London. Glass urns and bottles, as well as bowls, cups, and dishes of glass are of considerable value on account of their form and colour,

and the pottery of various wares makes perhaps the most
striking appeal to a visitor to the Guildhall Museum.
Here may be seen urns, vases, bowls, and cups of Samian
and Upchurch wares, which have been dug out in Aldgate,
London Wall, Bishopsgate, Lombard Street, and other

London Stone, Cannon Street
(*From an old print*)

places. In 1677, on the site of the present St Paul's, a
Roman kiln was discovered, where coarse pottery as well
as all kinds of tiles were manufactured.

The Romans had burial-places and tombs in Bow
Lane, Camomile Street, Cornhill, and St Paul's Churchyard, and sarcophagi of stone and marble have been found

at Clapton and the Minories. Leaden coffins were discovered at Old Ford in 1844, and at Bethnal Green in 1862, and monumental stones and slabs are sometimes unearthed.

There is one other relic of Roman London to which we must briefly refer. London Stone is built into the south wall of St Swithin's Church in Cannon Street. It is probably an old Roman mile-stone, which may have marked the beginning of the first mile on the Watling Street. London Stone is one of the most valued relics of London, and in the Middle Ages it was very greatly esteemed.

When we pass from the Roman period to the time of the Early English we find a positive dearth of antiquities to represent this later age. Besant says "there is nothing, absolutely not one single stone, to illustrate Saxon London," but he thinks that some of the columns in Westminster Abbey and the Chapel of the Pyx represent the work of the Confessor. In 1774 an earthen vessel containing coins of Edward, Harold, and William was found near St Mary-at-Hill, and an enamelled ouche in gold of the ninth century was discovered near Dowgate Hill. Personal ornaments and requisites of metal, bone, and horn, as well as weapons and tools of the Early English period, have been dug up in various parts, but they are not of sufficient importance to be specified.

What has survived to remind us of London of this date is of more importance perhaps than many relics. The names of many streets such as Cheapside, Ludgate, Bishopsgate, Coleman Street, Cornhill, Walbrook, and

Gracechurch Street are without doubt of Early English origin; and the churches dedicated to St Botolph, St Ethelburga, and All Hallows remind us of some of the early saints. Again, the same influence is evident in the usage to-day of such words as alderman and sheriff, and of the name of the meeting-place of the councillors—the Guildhall.

18. Architecture. (a) Ecclesiastical— Medieval Churches. Wren's City Churches.

When we consider the ecclesiastical architecture of London, we find that it occupies quite a different position from that of the other English counties. In these, as we know, there are hundreds of old churches, some dating from Norman and even earlier times, but in London the majority of the churches are modern. This fact, of course, is largely owing to the Great Fire of 1666, which swept away some of the finest churches of the earlier periods.

Before we notice the results of the Great Fire, it will be well to glance at the ecclesiastical condition of London prior to 1666. Perhaps the first thing that strikes us is the large number of parishes, each with its own church, within the walls of ancient London. Fitzstephen tells us that in his time there were 13 large conventual churches and 126 lesser parochial churches; and later Stow, the great historian of London, gives a list of 125 churches, including St Paul's and Westminster Abbey.

At the present time the City of London has many
fine churches ; and from the Surrey side of the Thames
one is struck with the lofty steeples and spires that tower
in their beauty above the warehouses and places of
business. Let us think for a moment what place the
church occupied in the medieval life of London.   Then
the resident population of the City was much greater than
it is now, and many of the parishes were very small.   At
every street corner rose a church, and the City was filled
with priests, friars, and other ministers of the church.
While some of the churches were small, many were rich
and costly buildings, which had been beautified and
adorned by the loving thought of many generations.

The church then occupied a large part of the daily
life of the people, who were expected to attend the many
services in their parish churches.   The gilds and trade
companies went to church in state.   There were no
sectarian differences as now.   All the people belonged to
the one Church, and its services were part of their daily
life, as much as their work, their food, and their rest.

We need not go into details as to the changes that
were brought about by the Reformation, or to the ravages
of the Great Fire.   We may note, however, in passing
that 89 churches were destroyed in 1666, and of them
only 45 were rebuilt.   Among the churches that escaped
the Great Fire were St Giles's Cripplegate, St Helen's
Bishopsgate, St Bartholomew's Smithfield, St Katherine
Cree Leadenhall Street, All Hallows Barking, St Olave's
Hart Street, St Etheldreda's Ely Place, St Ethelburga's
Bishopsgate Street, St Andrew Undershaft, and the

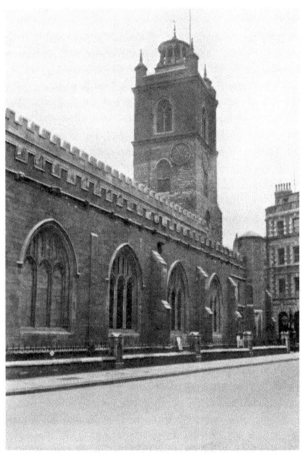

St Giles's, Cripplegate

Church of the Austin Friars. Pepys the diarist has the following observation on the subject of the London churches destroyed in the Great Fire :—" It is observed and is true in the late Fire of London, that the fire burned just as many parish churches as there were hours from the beginning to the end of the Fire ; and next, that there were just as many churches left standing as there were taverns left standing in the rest of the City that was not burned, being, I think, thirteen in all of each, which is pretty to observe."

There are, however, remains of early and medieval churches in London which are worth considering and are representative of all styles of English architecture. The successive periods of English church architecture are generally given as Norman or Romanesque, Early English, Decorated, and Perpendicular, and range in date from the eleventh to the sixteenth century.

The Norman period of church architecture was from 1070 to 1154 ; and of early Norman work London has some good examples in the chapel of St John in the White Tower, and the crypt under the Church of St Mary-le-Bow, Cheapside. The choir of St Bartholomew, Smithfield, and some details in the nave of St Saviour's Cathedral, Southwark, illustrate later Norman work.

Towards the end of the twelfth century the round arches and heavy columns of Norman work began gradually to give place to the pointed arch and lighter style of the period of English church architecture known as Early English, which is so conspicuous for its long narrow windows. The choir and eastern chapels of

St Saviour's Cathedral and Lambeth Palace Chapel exhibit the Early English style at its best. The transepts and choir of Westminster Abbey are most beautiful examples of Early English, but of this building, which illustrates English Gothic architecture in all its phases, we must defer our consideration for the volume on West London.

The Nave, St Saviour's Cathedral

The Early English style flourished from 1154 to 1270 and gave place to the highest development of Gothic— the Decorated, which prevailed throughout the greater part of the fourteenth century, and was particularly characterised by its elaborate window tracery. The chapel of St Etheldreda in Ely Place, Holborn, and the lower chapel of St Stephen in the Houses of Parliament,

Westminster Abbey

illustrate the Early Decorated, while the windows in the nave of the Austin Friars' Church, and the south transept of St Saviour's Cathedral give a good idea of the work of the Late Decorated period.

The Perpendicular is the name given to the last period of Gothic architecture in England; it established itself from the Decorated towards the end of the fourteenth century and was in use till about the middle of the sixteenth century. The Perpendicular, which, as its name implies, is remarkable for the perpendicular arrangement of the tracery, and also for the flattened arches and the square arrangement of the mouldings over them, is seen only in England. It is to the Perpendicular period that the old churches situated in the northern and eastern parts of the City chiefly belong. They are St Giles's Cripplegate, St Helen's and St Ethelburga's Bishopsgate, St Andrew Undershaft, St Olave's Hart Street, All Hallows Barking, and St Peter ad Vincula within the Tower. The old churches at Hackney, Stoke Newington, and Hornsey also date from this period.

There is not space to go into details with regard to the old churches of the parishes outside the City of London. As a rule they are not elaborate, and offer few examples of artistic detail. In no way do they compare favourably with those of Essex, Suffolk, or Norfolk, either in point of size or beauty of parts. Plainness and simplicity are the characteristics of these churches; and the plans are almost always a nave and chancel, a south porch, and a western tower. Their architects had to build with the materials they could command; and as these were different from what are found elsewhere, these churches are unlike those in the City of London, where the most expensive materials were used in the fabrics of the fine buildings within the walls.

It will be recognised that the previous remarks apply only to those churches existing in London prior to the Reformation.   From that period to the Great Fire, there

St Saviour's Cathedral, Southwark

was little church building, and that such as to require no comment in this chapter.   When, however, the Great Fire wrought such havoc, it was decided to erect fifty-three new churches upon the sites of those burnt, or so much damaged as to require rebuilding.   Sir Christopher Wren was the great architect to whom was entrusted this work of church building and restoration, and it is

St Dunstan's-in-the-East

generally acknowledged that the results were excellent. A recent writer says :—" Nothing that has been achieved in modern architecture has surpassed the beauty of their steeples, not only from the elegance of each, but for their complete variety, while at the same time in harmony with one another.   No two are alike.   The view of

St Clement Danes

the City of London from the old Blackfriars Bridge— up to the middle of the last century—must have been scarcely surpassed in any country : and all this was the work of one man ! "

Among the best of Wren's churches in the eastern part of London, the following are the most noteworthy :

St Bride's Fleet Street, St Michael's Cornhill, St Stephen's
Walbrook, St Lawrence Jewry, St Dunstan's-in-the-East,
St Swithin's Cannon Street, and St Magnus' London
Bridge. Wren's churches are remarkable for the variety
and originality shown in their design ; their elegant
proportions are noteworthy and satisfactory ; and it may
safely be said that no subsequent English church architect
has approached the fine work of Wren.

St Stephen's, Walbrook

Reference has already been made to the handsome
steeples of Wren's churches, but those of Bow Church,
Cheapside, and St Bride's, Fleet Street, are the most
admirable examples. St Bride's steeple is one of the finest
creations of Wren's genius ; it is of quite unusual type,
showing a series of six octagons diminishing as they
ascend.

The interiors of Wren's churches are also successful, exhibiting black and white pavements of marble, and much beautiful carving in the pulpits, pews, and organ cases. Wren did not use the Gothic style that had been in vogue before his time, and all his churches were based on the Classic. Since his day a reaction has set in, and during the last fifty years most of the London churches have been constructed on medieval ideas. Wren's churches, however, are works of architecture the like of which are possessed by no other place, and they are worthy of the highest reverence.

19.   Architecture.   (b)   Ecclesiastical— St Paul's Cathedral.

We briefly considered in the previous chapter the great skill of Wren as an architect, for in the fifty-three churches which he was commissioned to erect after the Great Fire, he showed an inexhaustible fertility of invention, combined with good taste and a profound knowledge of the principles of his art. We shall therefore do well in this chapter to pass on to a review of St Paul's Cathedral, which, without doubt, is Wren's greatest achievement. The steeples of Wren's churches are the most striking features in the view of London from the Surrey side, but the dome of St Paul's dominates the whole City. Owing to the neighbouring warehouses and shops, it is difficult to get a good view of St Paul's, but when looked at from Black-friars Bridge it is unspeakably grand and noble, and

St Paul's Cathedral from the S.W.

is the central object and most prominent feature of modern London.

The present Cathedral of St Paul is the third that has stood upon this site. The first cathedral is said to have been built by King Ethelbert in 610. We know little about it, but it was burnt down in the reign of William the Conqueror. In 1087 a second cathedral was begun, and this great building took nearly two hundred years to complete. In our history it is known as Old St Paul's. The building was 590 feet long, and the spire 520 feet in height. Near the cathedral stood the celebrated Cross of St Paul, where sermons were regularly preached. In the Great Fire of London in 1666, Old St Paul's was completely destroyed, and then it was that Sir Christopher Wren was entrusted with the work of planning and building the present cathedral, which is the finest building of the City.

St Paul's Cathedral as we know it to-day was begun in 1675 and completed in 1710. Wren lived long enough to enjoy the completion of his work, and there is a pretty story of how the architect in his great age, when he was too feeble to walk, used to be carried out to enjoy the sight of his fine building. The cathedral is indeed his monument, and over the door on the inside of the south transept is a tablet bearing this inscription in Latin:— " Si monumentum requiris, circumspice" (If you seek his monument, look around).

Old St Paul's with its lofty spire, its pointed windows, and lofty, narrow arches was quite different from the modern St Paul's with the dome and square or round-topped windows. The former building was a magnificent triumph of Gothic architecture, whereas the present cathedral is a fine example of the Classical style, showing, as it does, more than one order of Greek architecture.

The church resembles St Peter's at Rome, though on a smaller scale, and is in the form of a Latin cross. It is 500 feet in length, and 118 feet wide. The dome is 102 feet in diameter, and the top of the cross which surmounts the dome is 404 feet above the ground. St Paul's is the third largest church in Europe, being surpassed only by St Peter's at Rome and the Cathedral of Milan.

The grand entrance of St Paul's is in the western portico facing Ludgate Hill, and is reached by a flight of marble steps. This western façade has Corinthian pillars in the lower part, while those in the upper portion are Composite. On the apex of the pediment is a colossal figure of St Paul, and right and left are statues of St Peter and St James. On each side of the facade is a campanile tower, with statues of the four evangelists at the angles.

When we survey the interior from the principal entrance we are able to form a good idea of the length of the nave. The interior is somewhat cold and bare, and although much has been done of late years to relieve the great stretches of stonework, it lacks the warmth and colour of St Peter's at Rome. The dome was painted by Thornhill, and Sir William Richmond has recently decorated the roof of the choir with coloured mosaic. The reredos is a costly and sumptuous structure of white Parian marble, and in the nave some pictures by Watts and Holman Hunt have been recently placed on the pillars.

St Paul's Cathedral—West front

It is not necessary to describe in detail the many features of St Paul's, but it is well to remember that it is one of the two great burying-places of our nation. Now that Westminster Abbey is nearly full, St Paul's will be used as the place of sepulture for our worthies in the days that are to come. Some of England's greatest sailors lie buried in St Paul's. Here lie the remains of Nelson, enclosed in a wooden coffin made from the mainmast of the French flagship, *l'Orient*, which was captured at the battle of the Nile. Among other sailors of renown, Collingwood, Duncan, and Howe also remind us of the triumphs of our navy.

Wellington was buried in St Paul's, and his monument by Stevens is one of the finest in the cathedral. Sir Henry Lawrence of Lucknow renown, Sir John Moore the hero of Corunna, Lord Heathfield, who held Gibraltar for England, and General Gordon, who died at Khartum, are a few of the brave soldiers who are remembered at St Paul's by suitable memorials.

Not only are sailors and soldiers interred at St Paul's, or commemorated by stately monuments, but men of letters like Dr Johnson and Hallam the historian, philanthropists like John Howard, and artists like Sir Joshua Reynolds and Lord Leighton have also memorials in this our national Pantheon. The numerous monuments in St Paul's are not remarkable for their artistic merit, but they have a deep interest for all Englishmen and remind us that

> " Not once or twice in our rough island story
>   The path of Duty was the way to Glory."

The crypt is the most interesting portion of St Paul's, for in it are gathered nearly all the tombs that escaped the fire in Old St Paul's.   Here was buried Sir Christopher Wren, and here are the sarcophagi of Nelson and Wellington.   At the west end of the crypt is the hearse used at Wellington's funeral, made from guns captured by the Duke.

The Nelson Monument

St Paul's is famous for its fine peal of twelve bells which were hung in the north campanile tower in 1878. The south campanile tower has " Great Paul," the largest bell in England.   It was hung in 1882, and weighs more than 16 tons.

The cathedral has been the centre of our religious life for two centuries.   Its bells have rung for public rejoicings and national thanksgivings, and its great bell has sounded the passing of great Englishmen. Perhaps the most memorable service was that in 1897 when Queen Victoria returned thanks to Almighty God on the occasion of her Diamond Jubilee. That service was held outside the cathedral, and on the spot where Queen Victoria's carriage stood an inscription has been cut in the granite recording this noteworthy event.

20. Architecture. (c) Ecclesiastical— The Religious Houses.

From the previous chapters on the churches of London it will be gathered that the influence of the Church was very great during the medieval period. Not only were there 125 or more churches within the London of that time, but there were also at least twenty religious houses either within or just without the City wall. The religious houses were possessed of great wealth, and besides the monks, friars, and nuns, they retained a host of officers and servants.

Now let us consider the parts of a great monastery. The church of course occupied the chief place, for it was the centre of the regular life; it was generally situated on the north side of the monastic buildings. Next in importance came the cloisters, whose four walks surrounding the garth formed the dwelling-place of the community. The chapter-house was on the east side of the cloisters, and

Paul's Cross

Paul's Cross was a pulpit cross of timber, mounted on steps of stone
and covered with a conical roof of lead. Sermons were preached
from it by learned divines every Sunday morning. It was pulled
down by order of Parliament in 1643.

the refectory, or common hall for all conventual meals, was generally placed as far as possible from the church. The kitchen and other offices were near the refectory, and the dormitory contained the cubicles or cells, for each monk had a little chamber to himself. The other chief parts of a monastery were the infirmary, or house for the sick and aged; the guest-house, always open to give hospitality to strangers; the parlour, or place of business; the almonry, where the poor could come and beg; the common room, where the monks warmed themselves in winter; and the library, where the books and manuscripts were kept. From this summary it will be readily understood that the religious houses occupied a very large part of medieval London, and although so little of them is left to-day, their importance before the Reformation must not be forgotten.

The religious houses of London were divided among various Orders of monks, friars, and nuns, and it is interesting to note that many of the streets of the City to-day retain some of the old names which were originally given when these Houses were formed. We have space sufficient only to glance at a few of the chief Orders. The Benedictines are the most ancient and the most learned Order, and their House in London was known as the Priory of the Holy Trinity, Aldgate. It was a House of the first importance in London, and the Pope absolved it from all jurisdiction. Springing from this Order were the Carthusians, who occupied the Charter House in London, and the Cistercians, who had Eastminster, or the Abbey of St Mary of Grace.

The Temple Church

The Austin Friars, said to have been founded by St Augustine, had a monastery, and their name is still used to describe a Court in Old Broad Street.

The three great Orders of mendicant friars, the Franciscans, Dominicans, and Carmelites, were very popular in London. The Franciscans, founded by St Francis of Assisi, had Grey Friars House, which latterly became Christ's Hospital, familiarly the Blue-coat School. The Dominicans or Blackfriars, founded by St Dominic, established themselves in the locality still known by their name. The Carmelites were the Whitefriars, whose House was in Fleet Street. The district around their monastery is even now known as Whitefriars, and for a long period it had the privilege of sanctuary, which was not abolished till 1697.

Among the other Orders represented in London were the Cluniacs, who founded the Abbey of Bermondsey ; the Black Canons who were at St Bartholomew's, Smithfield ; and the Canons Regular of St Augustine who had the Southwark Priory of St Mary Overie, which is now the Cathedral Church of Southwark.

There were also two great military Orders which established themselves in London—the Knights Hospitallers of St John at Clerkenwell, where a fine gateway may still be seen (p. 167), and the Templars, who have left their beautiful church, the Temple Church off Fleet Street, to form one of the most interesting architectural relics of a bygone age. The Temple Church is one of four round churches in England, and has some monuments of Templars of the twelfth and thirteenth centuries. They are recumbent figures of dark marble in full armour, and are most beautifully executed.

The Church of St Helen in Bishopsgate Street reminds us of the Priory that once stood there. This church escaped the Great Fire, and, on account of its importance as a place of sepulture, it is often called the Westminster Abbey of the City.

St Bartholomew the Great, Smithfield

The Church of St Bartholomew, to which reference has already been made, encloses the choir and transepts of the ancient Priory Church of St Bartholomew, and is the most interesting church in London. Its founder was Rahere, a favourite and witty courtier of Henry I, and the monastic buildings were of great extent. The interior of the church has some striking Norman work in the choir, and also in parts of the nave and transepts. The special interest of the church centres around the tomb of the founder. The recumbent effigy of Rahere, the first prior, is under a rich vaulted canopy, the work of the fifteenth century. The crypt, recently excavated, is well preserved, and the ancient churchyard opens into Smithfield through

Interior of St Bartholomew the Great

the old Norman arch. Close by is St Bartholomew's Hospital, which is also a foundation of Rahere.

Such then were some of the religious houses in London before their dissolution in Henry VIII's reign. So full was London of these foundations that, after they had fallen, large portions of the town were left desolate. Most of the relics of the various monasteries which once occupied so large a part of London have been entirely swept away, and only their names survive, either in the localities where they stood, or in some later scholastic or hospital foundation.

It is generally admitted that most of the religious houses were seats of learning, and if it had not been for the monks most of the arts, and much of the science and scholarship of our country would have perished.   Besides all this, the monks offered an asylum for the poor and oppressed, and in an age of unrestrained passion they showed an example of self-restraint and poverty.   The monks, too, did much practical work, for the Thames was embanked at Bermondsey and Rotherhithe by the monks of Bermondsey Abbey, and thus those two low-lying districts were saved from floods.   However, Henry VIII was persuaded to lay violent hands on the religious houses of England and Wales, and this no doubt was with a view to enrich himself and to reward his courtiers. His harsh action told very heavily on the poor, whose needs had to be considered and provided for in some other way.

21.   Architecture.   (*d*)   Military—The Walls and Gates.   The Tower.

The need for the protection of London was recognised from the earliest years.   We need not stay to consider the theories with regard to the defensive work of the Britons, but we will pass at once to the Roman period.

It is now generally accepted that the first Roman fort
was founded by Aulus Plautius in the year 43, and during
recent years its site has been examined. It stood above
the Walbrook, and extended from the present Cannon
Street Station on the west to Mincing Lane on the east.
This Roman fort covered an area of about 50 acres, and
was the barracks, the arsenal, and the treasury of the
station.   It was the official residence, and also contained
the law-courts and the prisons.

The Romans, however, depended on their walls and
their arms rather than on the fortified position of the
town.   Hence we may fairly consider the Wall of
London as one of the principal Roman buildings, and we
shall not be far wrong in assigning the years between
350 and 369 for its erection. The complete circuit of
the wall can still be accurately traced from various existing remnants, and from old plans and records. It is
certainly remarkable that for more than fifteen hundred
years the boundary of the City was determined by this
wall, and that for some purposes the wall still affects the
government of London.

This remarkable wall was about 10 or 12 feet thick,
and probably from 20 to 25 feet in height. It was
composed of rubble and mortar, and faced with stone.
Roman work may be known by the courses of tiles or
bricks, which are arranged in double layers about 2 feet
apart. It may be mentioned that the so-called bricks are
not at all like our bricks, for they generally measure
18 inches in length, 12 inches in width, and $1\frac{1}{2}$ inches in
thickness.   At intervals the wall was strengthened with

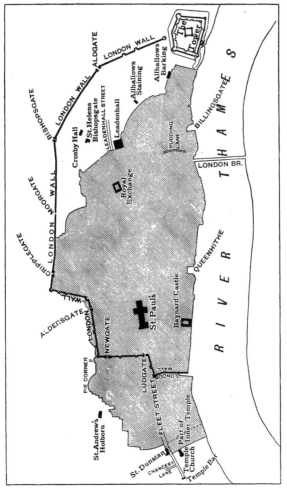

Plan of Old London : showing the Wall and Gates

(The shaded area was that destroyed by the Great Fire)

Remains of London Wall near Postern Row, Tower Hill, in 1818

towers and bastions about 50 feet high, and afterwards there was also a moat or ditch outside the enclosure.

Fitzstephen, who wrote about London in the twelfth century, describes it as "the great and high wall of the city having seven double gates and towered to the north at intervals; it was walled and towered in like manner on the south, but the Thames has thrown down those walls." The wall had a walk all round, a parapet, and battlements, and part of the wall which still shows the walk and battlements may yet be seen in London Wall, while another part of the wall is in the churchyard of St Giles's, Cripplegate.

We need not go into details with regard to its later history, but after the departure of the Romans it fell into a ruinous condition, and it was necessary for King Alfred to rebuild and strengthen it. Other repairs were made at various times from the reign of King John to that of Edward IV, but we must always remember that the course of the original wall was never altered. The total length of the wall was about $3\frac{1}{8}$ miles, and it enclosed an area of 380 acres.

Now, having considered the nature of the Roman defence of London, we may briefly trace the course of this formidable wall. If we begin in the east where the Tower now stands, we shall find that it followed a course nearly north to Aldgate and then turned off north-west to Bishopsgate. Thence it continued, past what is still called London Wall, to Monkwell Street, where it turned south till Aldersgate was reached. From here it ran west to Newgate, and at that point turned due south to the

Ald Gate

Bishops Gate

Moor Gate

Cripple Gate

The Gates of London
(*From old prints*)

Alders Gate   New Gate

Lud Gate   Temple Bar

The Gates of London and Temple Bar
(*From old prints*)

river at Blackfriars. Thence its course was along the river bank to join its extremity at the Tower. In the river wall there were at least two openings or gates which were known as Dowgate and Queenhithe.

The gates of the City were finally removed in 1760, and by that year most of the wall was gone. Besides those portions of the wall already mentioned, others have been discovered where excavations have been made, and there is no doubt that the foundations still exist from one end to the other of this remarkable Roman construction.

So much for the walls and gates of London; we may now pass to the Tower of London, which is the most celebrated fortress in Great Britain, and historically the most interesting building in London. It has existed for over eight centuries, and tradition even ascribes it to Roman foundation. It stands just without the City walls, on the left or Middlesex bank of the Thames, and "below bridge." Stow, the great historian of London in the sixteenth century, thus describes it :—"The Tower is a citadel to defend or command the City ; a royal palace for assemblies or treaties ; a prison of state for the most dangerous offenders ; the only place of coinage for all England ·at this time ; the armoury for warlike provisions ; the treasury of the ornaments and jewels of the Crown ; and general conserver of the King's courts of justice at Westminster." Through the course of centuries it has been in turns a fortress, a palace, and a prison, while to-day it is a show place carefully kept up, and utilised as a soldiers' barracks and a government arsenal.

Perhaps we may still consider the Tower as a fortress, but, before we review its history and architecture, we will pause a moment to glance at a quaint ceremony that takes place each night, after the gates are locked, and no one can enter without the password, which is changed daily. At a few minutes before eleven, the yeoman porter takes his keys and asks the serjeant for

St John's Gate, Clerkenwell

the "Escort for the keys." The serjeant informs the officer, who, placing the guard under arms, furnishes a serjeant and four men. Two of the men are unarmed, for they must close the gates, and carry the ancient lantern, which burns a tallow candle. The procession is then formed, and in the midst of the escort is the yeoman porter with the keys. He goes the round of the gates,

The Tower of London

and on his return to the main guard the sentry at the guard-room exclaims :—" Halt ! who comes there ? " to which the yeoman porter replies " The keys." Another challenge follows—" Whose keys ? " and the answer is, " King George's keys." Then the yeoman porter hearing the order, " Advance King George's keys," places himself in front of the guard. The guard present arms and the yeoman porter says, " God preserve King George," to which comes the response from all, "Amen." This daily incident links the past with the present, and helps us to realise the continuity of our history.

The ground plan of the Tower is in the form of an irregular pentagon, which encloses an area of 13 acres, and is surrounded by a double line of circumvallation —the outer ballium or bailey and the inner ballium— strengthened with towers. In the midst rises the square White Tower, which was built by Gundulf, Bishop of Rochester, in 1078. The work of building the Tower took many years to complete, and although the Conqueror planned this great fortress, he did not build more than the inner ward. The Tower was further strengthened by William Rufus, and Henry III made extensive alterations and additions to it, especially the outer ward, which completely surrounds the inner ward, and is itself contained within the bridge or moat. The moat was drained and made a garden, as we now see it, in 1843.

There are four entrances to the Tower, those on the Thames side being the Iron Gate, the Water Gate, and the Traitors' Gate, but the latter is now disused. The principal entrance on the west is the Lions' Gate, which is named from the menagerie of lions once kept in the Tower. The Lions' Gate is under the Middle Tower, and is defended by a portcullis. A stone bridge over the moat leads to the outer bailey, and here we see Traitors' Gate, the former entrance of state prisoners who were brought hither by water. It is

> "That gate misnamed, through which before
> Went Sidney, Russell, Ralegh, Cranmer, More."

A gateway opposite the Traitors' Gate leads under the Bloody Tower, which was the scene of the murder of the young princes by command of their uncle, Richard III, and we pass into the Inner Bailey. This is a wide, flagged courtyard, where the soldiers are drilled, and in the centre rises the fine keep of the castle, which is now called the White Tower.

As we have already remarked, the White Tower is the most ancient part of the fortress. It is an immense square building with corner turrets, and pierced with Norman windows and arches. It has four tiers—the vaults, the main floor, the banqueting floor, and the state floor. The height of this keep is 92 feet, and the walls are from 13 to 15 feet in thickness. It was under the staircase of the White Tower that the bones of the murdered princes were found. The Chapel of St John is on the second floor, and with its massive pillars and cubical capitals is one of the best specimens of Norman architecture in England. A collection of old armour of great value and interest is now kept in the two upper floors of the White Tower.

Outside the White Tower is an interesting collection of old cannon, some dating from the time of Henry VIII. The Record or Wakefield Tower contains the Crown Jewels, which are kept in a large vaulted chamber, under the strictest guard, for their value is estimated at £3,000,000. A glance at a plan of the Tower will show that altogether there are twelve towers of the Inner

The White Tower

Ward, and these were all used at one time as prisons, and held some most notable prisoners. Among the most celebrated may be named John Baliol, William Wallace, King Henry VI, Cranmer, Sir Thomas Wyatt, Sir Walter Ralegh, Earl of Strafford, Archbishop Laud, Lord William Russell, the Seven Bishops, and the infamous Jeffreys.

The Tower was not only a prison, but it was also a place of execution. Tower Green is, indeed, a spot of hallowed memories, for among those who were beheaded here were Anne Boleyn, the Countess of Salisbury, Lady

Sir Thomas More

Jane Grey, and the Earl of Essex. The place of execution is now a gravelled enclosure, which was railed in by command of Queen Victoria. Many other eminent persons, such as Sir Thomas More, were executed on

Tower Hill, which was under the authority of the Governors of the City.

The last building we must mention within the Tower is the little chapel of St Peter ad Vincula, the Prisoners' Chapel, dating from 1305. It is chiefly interesting from the fact that it is the burial-place of so many distinguished persons who lost their lives in the Tower. On the chapel door is a memorial tablet containing the names of thirty-four famous persons buried in it. Macaulay writing of this chapel says :—" In truth there is no sadder spot on earth than this little cemetery. Hither have been carried through successive ages, by the rude hands of gaolers, without one mourner following, the bleeding relics of men who have been the captains of armies, the leaders of parties, the oracles of senates, and the ornaments of courts."

22. Architecture. (e) Domestic—Palaces, Houses, Halls of City Companies.

London was once a city of splendid palaces and houses, nearly all of which were destroyed by the Great Fire. We have also seen that the City was covered with magnificent buildings of monasteries and churches, and it is not at all difficult for us to picture to ourselves what medieval London was like. The streets were very narrow, and there was no attempt to group the fine houses of the nobles and rich merchants, with their spacious

courts and gardens.   There is very little to remind us of
those times, and we must form our ideas from old prints
and descriptions of the City.   After the Great Fire,
Wren drew up a plan to rebuild London, and if he had
been able to carry his ideas into execution there would
have been a really fine city with broad streets.   Instead
of a well-planned city, however, the houses sprang up
on almost the same sites, and there remained nooks and
corners, quaint alleys and courts, many of which are still
evident.   London, however, changes so quickly, that
there are not many houses left of Wren's time.   Every
year the City becomes more and more a collection of
offices and warehouses, and even during the last few
years some historical houses have given place to buildings
that are utilitarian and not beautiful.

The great palaces of London have always been in the
West End of what we now call the County of London, so
that in this book we will not refer to them.   There are,
however, two royal palaces in this eastern portion of
London which were formerly of some note and may
therefore receive a passing notice.   The ancient palace
of Eltham dates from about 1300, and is interesting as
having been the residence of some of our English kings.
From the time of Henry III to that of Henry VIII it
was often the scene of royal splendour and feasting; and
its magnificent banqueting-hall must have been a very
suitable council chamber for the meetings connected with
state business.   This hall with its fine roof and beautiful
oriel and other windows, as well as the ancient bridge
which still spans the moat, dates from the fifteenth century.

Greenwich was also a royal residence as early as 1300, and several of our monarchs were born in the palace or lived there. Among others, Henry VIII, Queen Mary, and Queen Elizabeth were born at Greenwich Palace, while Edward VI died there. James I built a new brick front on the garden side, and afterwards it passed into the hands of Charles I, Cromwell, and

Greenwich Hospital

Charles II. The last-named monarch pulled down the old building, and erected part of a very fine palace, which Queen Mary afterwards converted into a hospital for seamen. This splendid building is now the Royal Naval College, and the Painted Hall is one of its chief points of interest.

Now leaving the palaces we may pass on to consider a few of the most interesting houses that formerly stood in the City, and also note the very few that are still standing. We have already remarked that London was a city of great houses and palaces, and it is probable that few cities in Europe could compare with it in the Middle Ages. Not only did London then have its merchants' palaces, but the town houses of the chief nobles of the land were within its borders. In the City alone, before the Great Fire, there were houses of the Earls of Arundel,

Baynard's Castle

Northumberland, Essex, Richmond, Warwick, and Westmorland, and also fine houses of the bishops and the abbots. These great houses of the nobles were always built in the form of a quadrangle, and of them the College of Arms in Queen Victoria Street is a good illustration, for it was Lord Derby's town house.

Baynard's Castle stood on the banks of the Thames just below St Paul's, "and was so called after Baynard, a nobleman that came in with the Conqueror." This fortress, with its towers and turrets, was rebuilt by Humphrey, Duke of Gloucester, and afterwards fell into the hands of Richard, Duke of York. Here it was that Edward IV assumed the Crown, and that Buckingham offered the Crown to Richard ; here it was, too, that Henry VIII lived, and Charles II was entertained. Shakespeare refers to Baynard's Castle in *Richard III*, and Pepys mentions it in his *Diary*. Baynard's Castle was destroyed in the Great Fire, but a memory of it is preserved in the name it has given to the Ward of Castle Baynard.

Less than forty years ago, some of the oldest houses in the City of London were to be seen in the Ward of Bishopsgate, for that was one of the districts that escaped the terrible devastation of the Great Fire. Many of those buildings were of considerable excellence and illustrated the best work of the Elizabethan period. In Bishopsgate Street there was a striking group of five gabled houses which remained much the same as when first erected. The houses were constructed of wood, and the foundations of them were of entire trees, or of trees simply halved. The walls, both external and internal, were wholly of timber, filled up with plaster. This mode of construction goes far to explain the rapidity of the Great Fire, and the completeness of the destruction which it involved. These houses were of three floors, the highest of which opened, by a door placed exactly in the centre of each gable, to a kind of gallery protected by a rail.

In the same street there also stood the finest house of

Sir Paul Pindar's House

the kind in London, the well-known house of Sir Paul Pindar. The front towards the street, with its gable, bay windows, and matchless panel work, was one of the best specimens of the Elizabethan period. It was begun in one of the closing years of the reign of Elizabeth, on the return of its owner from his residence in Italy, and it stood in all its beauty till it was pulled down to make way for Liverpool Street Station. The fine carved oak front of the house was then presented by the Great Eastern Railway to the South Kensington Museum, where it may still be seen.

Crosby Hall was the most interesting house that survived to our own day, for it was only pulled down and removed in 1908. It also was in Bishopsgate Street, and was a most beautiful specimen of fifteenth century domestic architecture. The only remaining building in London of its style, it was built by Sir John Crosby in 1481, and has figured largely in our history. It was here that Richard of Gloucester planned the death of his nephew, and in Shakespeare's *Richard III* there are several references to it as Crosby Place. The banqueting hall, with its fine timber roof, was very splendid ; and the oriel window had stained glass representing the armorial bearings of its various owners. When it was pulled down its materials were carefully preserved, and the building is now re-erected in Chelsea.

Another building of considerable historical and architectural interest is in Fleet Street. It was built in the reign of James I, and in " Prince Henry's Room " on the first floor there is some old panelling, as well as

Crosby Hall

a fine plaster ceiling. It has recently been carefully restored by the London County Council, and is now open to the public.

There are now very few of the old-fashioned roomy mansions left where merchants once lived close to their work, and there are not many modest dwellings now existent pleasing to the eye of the passer-by. There are still a few wooden houses in the locality known as Cloth Fair, but it is probable that they will soon make way for offices or shops. In Austin Friars and in Crosby Square are some fine old houses, which are worthy of notice. They are picturesque and stand in protest, as it were, against the change that is passing over London. So far as architecture is concerned it is not necessary to consider the streets and buildings of the area outside the City on the east side. The East End is almost a term of reproach, for its houses are devoid of style or interest.

We have referred to the City Companies in a former chapter, and here it will be fitting to note that most of them have Halls in the City. Some of these buildings are famous, and those belonging to the twelve great Companies are magnificent architecturally. The Goldsmiths' Hall in Foster Lane has a great pillared front. Its marble staircase is one of the most splendid in London, and its wide galleries are lined with costly marbles (see p. 95). The banqueting hall has portraits of royal personages, and the gold plate is magnificent. The Drapers' Hall in Throgmorton Street is built around a large quiet court. The exterior is plain and gives no idea of the splendour

"Prince Henry's Palace," Fleet Street

within.   The building has a handsome staircase of
coloured marbles, a fine banqueting hall, and statues and
pictures.   The Hall of the Merchant Taylors opens on
to Threadneedle Street, and is the largest of the Companies' Halls.   The great hall is a splendid room; its
windows are rich in stained glass, and its walls are
adorned by the arms of the members.   There are some
good pictures and many royal portraits, while the collection of gold plate is very valuable.   The small but
interesting crypt escaped the Great Fire.   What we have
said of these three Halls will apply in a greater or lesser
degree to those of such Companies as the Fishmongers,
the Stationers, the Grocers, the Mercers, the Salters, and
several others.

23. Communications — Ancient   and Modern.   The Thames formerly the Normal Highway of London. The Thames Watermen.

For many centuries the chief highway of London
was the Thames, which played a most important part
in the life and history of the City.   London has now
developed to such an extent that the number of people
who use the Thames either for business or pleasure is
really very small.   It is of the utmost importance, however, that we should realise that the Thames made
London, for it was the most important, if not the only

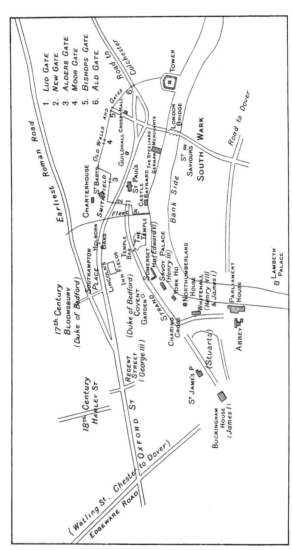

London and Westminster

Shewing the historical growth of the cities

highway by which merchandise in large quantities was brought into the City from the provinces, or from abroad.

The Thames, then, is the most natural starting-point when one considers the communications of London. A glance at a map of Roman London brings out clearly one important point. A great many of the ancient roads, both those of pre-Roman as well as of Roman days, seem to converge towards a single point on the northern bank of the Thames. Some of these roads, after traversing England for hundreds of miles in almost a straight line, are turned aside in order to reach that point. Now a little reflection will show the reason for this diversion of route. It was that the road might be carried over the ferry or bridge where the Thames was narrowest, and the present London Bridge is nearly on the site of the first ferry or bridge.

London owed its early prosperity to the building of the bridge, which is the first ascertained fact in the history of Roman London. We have already referred to the history of London Bridge in other chapters, so that it is not necessary to urge its importance all through our history. We are now in a position to consider, briefly, the roads through Roman London. The road from the south crossed the bridge to Eastcheap, where it divided into two branches, one of which ran northward to Bishopsgate, and the other north-westward to Newgate. The northward street at Bishopsgate again divided, the westward road ran to Lincoln and York, while the east-ward branch crossed the Lea at Old Ford and became the main road through Essex. The north-westward road

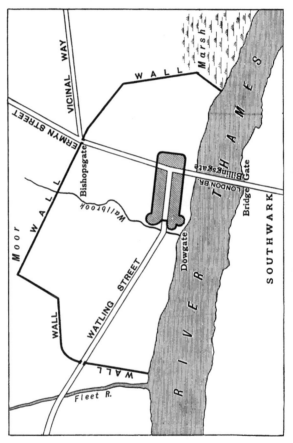

Roman Roads in Ancient London

passed from the City at Newgate, and throughout its entire length from Kent to its termination in Wales was known as Watling Street. One of the small streets in London, probably in the course of the original, still bears that name.

Besides the Thames, the Bridge, and the two or three main Roman roads, there was the Walbrook, a stream of some importance then, but which is now only a matter of history. Its name is retained by a thoroughfare by the Mansion House, and when excavations are made there are evidences of its former channel.

Now let us return for a short time to the Thames, which was for so long the normal highway of London. When the roads were few and bad, and when railways were unknown, it was considered safer and better to move from place to place by means of boats or barges. Londoners thus ran no risk of being stopped by footpads or highwaymen, and the "silent highway" of London was then used as much for pleasure as for business. It was not till the latter part of the nineteenth century that the Thames was embanked, and with a little thought we are able to realise that it was once broader than it is to-day. At high tide the water came up to the busy street we now call the Strand, which was then the strand or shore of the river. When the river was not embanked, it was not easy to get into a boat when the tide was low, and so in several places "stairs" were built which allowed persons to land or embark at all states of the tide. The popular old English song, "Wapping Old Stairs," reminds us of this fact. Water-gates, too,

were erected, and these allowed boats to come in at
any time and, from the little wharves which they contained, the owners of the gate could embark at any time

The Water Gate, Embankment Gardens

in their own barges. A good example of these water-
gates may still be seen at the bottom of Buckingham

Street, Strand.   This water-gate was designed by Inigo
Jones for the Duke of Buckingham, when he was living
at York House, his stately London home.

As the river was so much used by all classes, it is
evident that the number of boats must have been very
large.   Many of the citizens had their own boats, and
stately barges were kept by the Lord Mayor, the City

London Bridge

Companies, and the great nobles who then resided in their
town houses along the Thames.   We can form some idea
of the traffic on the river in the reign of Queen Elizabeth,
for Taylor the Water Poet tells us that in his time "the
number of watermen and those that lived and were
maintained by them, betwixt the bridge of Windsor and
Gravesend, could not be fewer than 40,000."   Later

we find that the watermen were made into a Company,
and that they could furnish 20,000 men for the fleet.
The watermen had the sole right to carry passengers for
hire upon the Thames, and were very zealous in pro-
tecting their rights. In 1850, a writer laments that
" the introduction of steamboats has changed the whole
character of the Company, and for every fifty watermen

Traitors' Gate, Tower of London

in the reign of Elizabeth, there is not more than one
now." We may go further and remark that practically
few steamboats now ply on the Thames, and passenger
traffic has almost ceased.

The Thames is of the greatest interest in our history,
and whole chapters might be written of the processions,
happy and unhappy, that have passed along its stream.

How many state prisoners have passed from their trials at Westminster to their doom at the Tower! Some of our greatest men have stopped in their boat outside the Tower, and, entering its gloomy portals by way of the Traitors' Gate, have gone to their cells, only to be beheaded after a short time on Tower Hill. The Thames carried the Seven Bishops to the Tower, and it became the repository for a time of the Great Seal of England, which James II in his flight threw into the water.

When Queen Elizabeth died at Richmond, her body was carried with great pomp by water to Whitehall, and a contemporary poet thus writes:—

> " The Queen was brought by water to Whitehall;
> At every stroke the oars did tears let fall."

Cowley the poet died at Chertsey and his body was borne by water to Whitehall, and Pope thus commemorates this event:—

> " Oh, early lost! what tears the river shed
> When the sad pomp along his banks were led."

Nor must we forget that a greater than Cowley· was brought in great state by water from Greenwich to Whitehall, for thus was Nelson carried to his last resting-place in 1805.

We must close our historical references to the Thames by a brief glance at Pepys. We cannot read much of the *Diary* without coming across such a phrase as, " By water to Woolwich," or " By water to Whitehall." The Thames plays a most important part in the London of Pepys, and right well did the diarist know how to amuse

himself by using the river in going from one place of
entertainment to another. Here are a few extracts from his
Diary on August 23, 1662:—"I walked all along Thames-
street but could not get a boat; I offered eight shillings
for a boat to attend me this afternoon, and they would
not, it being the day of the Queene's coming to town
from Hampton Court. So we fairly walked it to White
Hall, . . . and up to the top of the new Banqueting House
there, over the Thames, which was a most pleasant place
as any I could have got; and all the show consisted
chiefly in the number of boats and barges; and two
pageants, one of a King, and another of a Queene. . . .
Anon come the King and Queene in a barge under
a canopy with 1000 barges and boats I know, for we
could see no water for them, nor discern the King nor
Queene. And so they landed at White Hall Bridge, and
the great guns on the other side went off."

But now it is time to leave the Thames with its
merry-making, its pathos, and its tragedies, and pass to
the more prosaic study of the present-day communications
of London. The streets in the County of London are
maintained for the most part by the metropolitan borough
councils and the City Corporation. The London County
Council maintains the roadway of the county bridges, of
the Thames tunnels, and of the Victoria Embankment.
In the whole of the county of London there are 2135
miles of public roads and streets, and these are kept in
an excellent state of repair.

With regard to railway communications we find that
no fewer than ten trunk lines have their termini in

London, while local lines, which are now generally electrified, are numerous. The first electric railway in London was that from the Bank to Stockwell, which was opened in 1890, and the first of the "Tube" railways was the Central London from the Bank to Shepherd's Bush. The most frequented underground line is the

Liverpool Street Station

Inner Circle which is very convenient for travelling east to west, from Aldgate to Kensington. The best way to study the railways of London is to get a good railway map of the metropolis and trace the various lines. The total length of all the lines in London is about 250 miles. There are 329 stations in the County, and about 7800 trains enter and leave London every day.

Tramways have not yet been allowed to penetrate into the heart of London, but they are largely used in the northern and southern portions. Most of the tramway lines belong to the London County Council, and with few exceptions they are electrified. The chief starting-points of the trams for South London are from

Fleet Street, looking East

the Embankment; for North London, from Moorgate Street and the bottom of Gray's Inn Road; for East London from Bloomsbury and Aldgate ; and for North-West London from Tottenham Court Road. There are about 124 miles of tramways in London, and with the exception of a low-level underground tramway from

Theobald's Road to Aldwych, they are all above the surface.

The traffic of London is at present in a state of transition owing to the advent of motor 'buses. These are gradually superseding the horse-drawn vehicles, and although they were not introduced till 1899, they are now more than 1200 in number. The omnibus has long played a most important part in London locomotion, and from the top of a 'bus one gets some insight into the life of the City.

We now come to the last means of locomotion in London. The London cabs have taken the place of the old Hackney coaches and are now of three kinds—the four-wheeler, the two-wheeler, or "hansom," and the motor "taxi-cab." The latter class is rapidly increasing, while the "hansom" will soon be a thing of the past.

No city in the world is so well provided with the means of locomotion as London, and its history allows us to compare the state of affairs at various periods. London was famed for its coaching-houses till the advent of railways, and the coaches that ran from London to all parts of England were noted for their speed. In the reign of Charles II the fast coaches were called "Flying Machines," and the following is an advertisement recording what they could do :—"All those desirous to pass from London to Bath, or any other Place on their Road, let them repair to the Bell Savage on Ludgate Hill in London, where they may be received in a Stage Coach which performs the whole journey in Three Days (if God permit) and

sets forth at five in the Morning. Passengers to pay
One Pound five shillings each, who are allowed to carry
fourteen Pounds Weight—for all above to pay three
half-pence per Pound."

We have changed all that, and now we can leave
London and reach Bath in less than two hours. The
journey of 107 miles only costs 8s. 11d., and we have
no need to worry about attacks on the highway by
robbers or footpads.

24. Administration and Divisions — Ancient and Modern. The City Corporation. The London County Council.

London is the youngest of all our counties, and had
no central representative government till 1889, when the
first London County Council was constituted. From
1855 to 1888, the Metropolitan Board of Works was the
chief authority, but as it was not a popular body its work
was not altogether satisfactory. In this chapter we have
to consider the constitution and work of the London
County Council, which is the chief authority for the
Administrative County of London, and the constitution
and work of the Corporation of the City of London,
which has jurisdiction over a very small but most
important area.

As the City Corporation has a history that goes back for hundreds of years it will be well to begin with that body.   One of our great historians writes :—" London claims the first place as the greatest municipality, as the model on which, by their charters of liberties, the other large towns of the country were allowed or charged to adjust their usages, and as the most active, the most political, and the most ambitious."

We have not much knowledge of the government of the City of London before the Conquest, but probably it was similar to that which prevailed soon after that event. The most important event in its history after the Conquest was the granting of its first charter by William I, and this is addressed to the Bishop, as ecclesiastical governor, and to the Portreeve as civil governor. It is generally believed that there were portreeves until the first appointment of a Mayor, or Lord Mayor as he was styled in the sixteenth century.   A reeve was an officer appointed by the King, just as sheriffs, or shire-reeves, are still appointed.   The City of London obtained from Henry I the right of appointing their own sheriffs; and in the reign of King John the citizens gained their greatest privilege, for the Commune of London was formed.   The grant of the Commune involved a Mayor, and it was then the proud boast of London that "come what may, the Londoner shall have no king but their Mayor." When the City had obtained the privilege of the Commune and the right of electing their own Mayor, they chose Henry Fitzailwin for this position, and he continued in office until his death in 1212.

From that year to the present time London has elected its Mayor, or Lord Mayor, and many of these leaders of the City have been men of great character and authority. The way to the Mayoralty is by the Aldermanic Bench. Every Alderman, in course of time, becomes Lord Mayor, unless he resigns his position, or fails to secure election by the Court of Aldermen.

The Mansion House

The City Corporation of the present time is the representative of the Commune of seven hundred years ago. It is now styled "The Mayor, Commonalty, and Citizens of the City of London," and consists of the Lord Mayor, 25 Aldermen, and 206 Common Councillors. The Common Council for the City corresponds to the County Council for the whole of London. The

Common Council of the City of London is a democratic assembly, and the whole of its members are elected yearly by the ratepayers. Here we may mention that the Livery Companies are in close association with the municipal life of the City of London, for it is they who nominate the Lord Mayor and elect the two sheriffs on Midsummer Day.

The City of London Corporation extends its authority over Southwark and has most important powers as a municipality and as a judicial authority. The Corporation has its own police force, which is a most efficient body. It controls London, Blackfriars, Southwark, and Tower Bridges, and it is the Market Authority for London. The Corporation owns and maintains parks, open spaces, and Epping Forest; and it is a great educational authority, for it controls various schools, and the Guildhall Library and Museum. As a judicial authority the City has its Quarter Sessions; and the Lord Mayor and Aldermen sit at the Mansion House and Guildhall Justice Rooms, and also at the Central Criminal Court in the Old Bailey.

Among the chief officers of the Corporation are the Recorder, or the senior law officer; the City Chamberlain, or the treasurer and banker; the Town Clerk, who is the head of all the Corporation and Committee work; the Common Serjeant, who is the judge at the Central Criminal Court and the Mayor's Court; the Comptroller, who is the Conveyancing Officer; the City Remembrancer, who arranges the ceremonial functions; and the City Solicitor, who conducts legal proceedings and prosecutions.

Design for London County Council Offices

We now pass to a review of the constitution of the London County Council and its work. As already stated, it was formed in 1888, and began its duties in 1889. The Council consists of 118 elected representatives and 19 aldermen, and the election takes place in March, every three years. The aldermen are elected by the Council for six years, so that nine retire at the end of one period and ten at the next period. The first Chairman of the London County Council was Lord Rosebery, and other men of eminence have since been elected to this high position.

The work of the London County Council is of a most important character. One of the most interesting features of the Council is the care and development of the parks and open spaces, and of them some account is given in another chapter. It maintains the London Fire Brigade, which is one of the most efficient in the world; and it has important duties in connection with the health of the people, for it controls slaughter-houses, dairies, cow-sheds, and lodging-houses. The Council is constantly making improvements, by clearing insanitary areas, and by widening streets. It manages the lunatic asylums of London, and has a staff of officers to supervise weights and measures. The London County Council is the chief authority for education, and superseded the London School Board in 1903. Among the various other duties of the London County Council may be mentioned those relating to the management of the tramways, the housing of the working classes, and the preservation of historic buildings.

Besides the two chief governing bodies of London, the Administrative County has 28 Borough Councils, which by the Act of 1899 superseded the vestries and district boards. The council of each borough consists of a Mayor, and not more than 10 aldermen and 60 councillors. The powers and duties of these borough councils are not so important as those of the London County Council, but they are concerned with the maintenance of roads, and their cleansing and lighting; with public libraries, baths, and wash-houses, and other useful work within their area.

The County of London has 31 Boards of Guardians, four Boards of Managers of School Districts, and two Boards of Managers of Sick Asylum Districts. These various bodies are mainly concerned with poor law administration, with the management of workhouses, and with the work of caring for the poor, sick, and aged.

For the administration of justice, London has a Court of Quarter Sessions, and 15 Courts of Petty Sessions. There are also 14 Police Courts with Magistrates, and 14 County Courts. The Central Criminal Court has jurisdiction not only over all London and Middlesex, but also over parts of Essex, Kent, Surrey, and Hertford. There are prisons at Brixton, Holloway, Pentonville, Wandsworth, and Wormwood Scrubs, and the police force in London numbers about 18,000 men.

The County of London is in the dioceses of London, Southwark, St Albans, and Canterbury. The Archbishop of Canterbury has his official residence at Lambeth Palace, and the Bishop of London has Fulham Palace and London House, St James's Square. Formerly the ecclesiastical parish coincided with the civil parish, but now, while there are 69 civil parishes, there are 610 ecclesiastical parishes in London.

For parliamentary purposes London is divided into 58 constituencies, with one member for each, except in the case of the City, which returns two members.

The Metropolitan Water Board was formed in 1902, and has control of the water-supply, but its jurisdiction extends far beyond the County of London. The Port of London Authority was established in 1908, for the purpose of administering, preserving, and improving the Port of London. It superseded to a large extent the Thames Conservancy Board, which is now concerned mainly with the upper part of the Thames.

25. Public Buildings—Legal. Administrative. Hospitals.

In some of the preceding chapters we have referred to many of the important buildings of the eastern portion of London. There remain, however, some public buildings of considerable importance, and we will devote this chapter to a short account of them. Among the buildings connected with the law we will select the Temple, the home of the lawyers, and the Central Criminal Court in the Old Bailey ; among those dealing with the administrative work of the City and of the Government we may choose the Guildhall, the Mansion

House, and the General Post Office as typical buildings; and among the great hospitals, St Bartholomew's, Guy's, and the London are the best representatives.

The Inner and the Middle Temple occupy the whole of the large area between Fleet Street and the Victoria Embankment on one side, and Whitefriars to Essex Street, Strand, on the other. The Temple, as its name

Middle Temple Hall

implies, was originally the home of the religious and military Order of knights called the Knights Templar. They removed from Holborn to the present buildings in 1184, so the Temple may be said to be the most historically interesting of the Inns of Court. At first the Knights Templar were earnest and austere, but in course of time they degenerated, and were finally abolished by the Pope. In the fourteenth century the Temple was leased to the lawyers by the Crown, and in 1609 it became the free property of the Inner and Middle Temple. The original buildings of the Knights Templar have all disappeared, and in their places ranges of lofty buildings have arisen, with chambers for different tenants on every floor. The hall of the Inner Temple is fine and spacious, and the arms and crests of the treasurers of the Inn surround the hall, while on the walls are hung portraits of eminent lawyers and others. There is a good library. The famous Temple Church has been noticed elsewhere in this book. The Temple Gardens are mentioned by Shakespeare, who makes them the place where the red and white roses were plucked that were assumed as the badges of York and of Lancaster. Although the gardens no longer grow roses, they are noteworthy for their yearly shows of beautiful flowers. Nor must we forget the eminent men who lived and died in the Temple precincts. Oliver Goldsmith, Dr Johnson, and Charles Lamb are three names that occur among those who helped to shed lustre on this old-world place.

We now pass to a very different building, the Central Criminal Court, which occupies the site of the old Newgate prison. The Old Bailey, as it was called, stood like a medieval castle, square, rugged, and gloomy, and here till 1868 criminals were publicly hanged. In 1901, the place of execution was transferred to Holloway prison, and now the horrors of the Old Bailey are a matter of history. The new Central Criminal Court is a fine stone building, surmounted by a statue of Justice, holding

a sword in one hand and the scales in the other.   Above
the portal runs the inscription, " Defend the children of
the poor and punish the wrongdoer."   The present
building was erected by the City Corporation, and is
under their control.

Everyone acknowledges that the City is well governed,
and foreigners often express their delight at the clean and

The Guildhall

well-paved streets, and the efficient administration of the
Corporation.   The Guildhall is the Council Hall of the
City and stands in King Street to the north of Cheapside.
It dates from 1411, but the crypt, the old walls, and
the porch are all that remain of the ancient building,
which was much injured by the Great Fire.   The large
Hall has immense stained-glass windows, and a fine open

timber roof. The west end has the giant wooden figures of Gog and Magog, which formerly were used in the Lord Mayor's Show. The Hall is used for municipal meetings, for the election of the Lord Mayor, and for public banquets. It is full of historical memories, and in it there are statues, monuments, and portraits of

The General Post Office

great personages, such as Chatham, Pitt, Nelson, and Wellington. In connection with the Guildhall is the City Free Library, a very handsome modern Gothic building, with a valuable collection of books. The Museum is of considerable interest, as it contains a large number of antiquities found in London, and the Corporation Art Gallery has a good collection of modern British pictures.

The Mansion House, not far from the Guildhall, is the City residence of the Lord Mayor during his year of office. It was built by Dance in 1739, and the chief feature is the Egyptian Hall, so called because it was constructed on the model of the Egyptian Hall described by Vitruvius. At the Mansion House the Lord Mayor dispenses a lavish hospitality, and it is here that most of the great City banquets are held. Here also are held meetings of a philanthropic or benevolent character, and often the inauguration of a Mansion House Fund by the Lord Mayor results in the public subscription of a large sum of money.

Perhaps the most important Government buildings in the City are those known as the General Post Office. Besides the new buildings on part of the site of Christ's Hospital, there are three principal ranges of offices, known as *East*, *West*, and *North*. The old General Post Office East was superseded in 1910 by the New Building in King Edward Street, which gives improved accommodation for no less than 3000 sorters. The North building is the seat of the Accounts Staff, and the West building is used by the Telegraph and Engineer's Staff. The General Post Office was named St Martin's-le-Grand from an old collegiate church and sanctuary that stood on its site.

London is proud of its hospitals, and their administration is improving year by year. In no other city is so much done to alleviate pain and suffering. The subscriptions of all classes for the maintenance of the hospitals amount to a very large sum each year. There are three great funds which raise money for this purpose ;

there is King Edward's Hospital Fund, which was in-
augurated in the year of Queen Victoria's Diamond
Jubilee, and there are also the Hospital Sunday Fund,
and the Hospital Saturday Fund. Many of the large
hospitals are without medical schools, but some of the
oldest and best have them as part of the foundation.

St Bartholomew's Hospital

The total income of all the London hospitals in 1907
amounted to upwards of £800,000, and in the same
year more than 4,000,000 attendances were made by
out-patients, while nearly 40,000 in-patients were accom-
modated. Besides the large general hospitals, there are
special hospitals for the treatment of children's diseases,
consumption, skin and eye diseases, fever, and cancer.

St Bartholomew's Hospital was founded by Rahere, a favourite and courtier of Henry I, in 1123. It was refounded by Henry VIII, and is now the richest as well as the most ancient of the London hospitals. Besides attending to 300,000 out-patients in the year, it maintains 670 beds, and has a staff of 300 nurses. The main building was erected in 1733, but an extension has recently been added to it on part of the site of Christ's Hospital. This hospital is familiarly known as " Bart's " by the students, and is one of the very best of our schools for medicine and surgery. Harvey, the discoverer of the circulation of the blood, was one of its physicians, and Dr Abernethy used to lecture here.

Guy's Hospital in the Borough of Southwark was founded in 1721 by Thomas Guy, who was a bookseller and made a large fortune by printing and selling Bibles. His vast wealth went to build and endow this hospital, which is one of the largest of the London Medical Schools and maintains six hundred beds.

The London Hospital is in the Whitechapel Road, and was founded in 1740. It is by far the largest of all the London hospitals, for it has 914 beds, and the attendances of its out-patients yearly amount to nearly 600,000. Its endowment is not large, and so it depends mainly on subscriptions and donations. The building has no architectural pretensions, but it is of incalculable value to the teeming population of the East End.

26. Education—Primary, Secondary, and Technical. Foundation and Collegiate Schools. The University of London.

Before the year 1870, the elementary education of the children of London was not compulsory, and was managed by the Church of England and other religious denominations. Mr W. E. Forster introduced a bill for the compulsory attendance of all children at school, and when this Education Act was passed in 1870, a body known as the London School Board was formed. It consisted of 55 members, and for a period of 33 years was the directing authority for much of the elementary education in London. During its régime the Church, Roman Catholic, Wesleyan, and other denominational schools were controlled by their own managers, and had nothing to do with the London School Board.

The work of the London School Board came to an end in 1903, when Mr Balfour passed a new Education Act by which the London County Council became the Education Authority for the County of London. The London County Council actually superseded the London School Board on May 2, 1904, and an Education Committee was then formed to deal with all classes of education. At the present time the Education Committee consists of 40 members of the London County Council together with 12 co-opted members, who are specially chosen for their interest in the work of education.

The great merit of the last Education Act is due to the co-ordination of all branches of education in the hands of one body. Thus the London County Council have the charge of all the elementary schools, both those belonging to the late London School Board and the denominational schools. The latter schools, however, are still allowed to give their own religious instruction, and their managers have some control over the teachers in these schools.

Now, in considering the extensive duties of the London Education Committee we will begin with elementary education. There are about 920 schools for this purpose, and they have accommodation for over 750,000 scholars, whose ages vary from three to 15. The children have a sound elementary education, which is well graded for their capacities. There are also many special schools for instruction in such subjects as cookery, laundry work, housewifery, and manual work, and also for the separate treatment of children who are deaf, blind, and mentally or physically defective.

The Council have also 108 Higher Elementary Schools and Higher Grade Departments, where the curriculum is of a more advanced character, and the course of instruction is arranged for four years, after an entrance examination. For the whole of the elementary schools there are about 20,000 teachers employed, and altogether the London County Council expends upwards of three million pounds yearly on educational work.

The Higher Education of the London County Council began by taking over the duties of the Technical Board, and since 1904 it has been concerned with technical, secondary, and university education. The Council has adopted the policy of maintaining and developing the work of existing institutions in London before erecting new institutions under its own management. The most important of all the institutions which provide technical education are the Polytechnics; and the special aim of the Council has been to provide facilities for scientific and technological instruction in every district of London. At present there are 12 Polytechnics where instruction is given in the ordinary branches of science and art, as well as in the engineering, building, and chemical trades. There are other institutes which are specially devoted to the teaching of one particular craft, and these are styled Monotechnic Institutions. Thus there is one school devoted to training craftsmen in photo-process work and lithography, a second to carriage-building, and a third to leather-tanning and leather-dyeing.

The London County Council has numerous Schools of Art under its control, and there is also a Central School of Arts and Crafts, which provides instruction in decorative design for the artisans of London. The trades for which provision has been specially made at this school are those directly or indirectly associated with the building trades, such as decorators, stone-carvers, metal-workers, cabinet-makers, and designers of wall-papers.

Considerable grants are made by the London County Council to the University of London, and to University College, King's College, Bedford College, and the London School of Economics, and in return these bodies give

a certain number of free places to the nominees of the
Council. The London County Council spends large sums
of money in the award of scholarships which carry pupils
from elementary to secondary schools. It has now 16
secondary schools of its own, and it makes annual grants
to other schools which receive its scholarship holders.

Charterhouse

Besides the schools under the control of the London
County Council there are also some great Public Schools
which must be specially mentioned. First there is St
Paul's School, the most ancient, for it was founded in
1509 by Colet, Dean of St Paul's. For many years this
famous city school was under the shadow of the great
cathedral, but in 1884 it was removed to its present

fine buildings in the Hammersmith Road.   Westminster
School is at the back of Westminster Abbey ; it was
founded by Henry VIII out of the spoils of the monasteries, and richly endowed by Elizabeth.

Two other famous public schools formerly stood in
the very heart of the City.   Christ's Hospital, or as it is
familiarly called the " Blue-coat School," was one of the

Dulwich College

most cherished institutions of London.   It was founded
in the reign of Edward VI, and its scholars still wear the
picturesque dress of that period.   Owing to the want of
space, the school was removed to a more suitable site near
Horsham, in Sussex, and the old buildings have given
place to the extension of the General Post Office.   The
Charterhouse School was removed from London in 1872,
and was established in a fine building at Godalming,

in Surrey. The Charterhouse was founded in 1611 by Thomas Sutton, and though the boys have gone, the brethren of the foundation, some eighty in number, live on in the same place, in collegiate style. In addition to these great and famous schools of the past, there are others of a later date which give a similar education. Among these may be mentioned Dulwich College, the

City of London School

City of London School, University School, and King's College School.

We now come to the last section of this chapter, which has to deal with the University of London. Founded in 1836, the University was for many years an examining body, and had nothing to do with the work of teaching. Its examinations proved whether students had been well taught in certain subjects, and whether they merited its certificates or degrees. In 1898, however, the University became a teaching body as well as an examining board. Its headquarters were formerly at Burlington House, Piccadilly, but after its re-organisation, the central block and one of the main wings of the Imperial Institute at South Kensington were assigned for this purpose. The various colleges, and medical schools, such as King's College, University College, Bedford College and others, are now "Schools of the University." There are about 3500 students attending the 42 schools of the University.

27. Roll of Honour.

For some years before 1901, the Society of Arts placed tablets on houses of historical interest in London, but since that date the London County Council have undertaken this work. There are now more than one hundred houses thus indicated, but the work is by no means completed, for there are many celebrated men who have lived and died in London whose houses have not yet been marked by a memorial tablet. In addition to this commemoration of noted Londoners, there are also statues to some of them in the streets and the squares, as well as monuments to them in the cathedrals and churches of the metropolis. In this rapid survey of men who have conferred distinction on London, it will be possible to do little more than just mention the locality

with which they were connected, or the special work
which links them to the great City.   The Roll of Honour
of London necessarily includes men who are of world-
wide renown, and whose memory is honoured by other
towns and counties in our country, but there are many
of them who are indissolubly linked with the associations
of London.   To use the words of Lord Rosebery it may
be said that, in taking a walk in London, "it is an
immense relief to the thoughts to come on some tablet
which suggests a new train of thought, which may call to
your mind the career of some distinguished person, and
which takes off the intolerable pressure of the monotony
of endless streets."

We need not spend much time on the royal personages
who are associated more particularly with London.   The
fact that it has been the capital for nearly a thousand years,
and that the Court has resided there for the greater part
of that period, tells us at once that all our monarchs have
some claim, either by birth or residence, to be considered
Londoners.   There are a few, however, that we recall
at once for some special reasons.   Alfred has been called
the founder of London : William I built the Tower for
its protection ; and Charles I was beheaded at Whitehall.
Westminster Abbey with all its historic associations has
been the crowning-place of our sovereigns, and here, too,
many of them are buried.

Neither will it be necessary to recount the long list of
divines who have spent much of their time in London.
When we remember that St Paul's Cathedral and Westminster Abbey have given us a succession of noted bishops

and deans, and that Lambeth Palace, the official residence
of the Archbishop of Canterbury, has been the home of
nearly one hundred successors of St Augustine, we at
once realise what a part London has played in the religious life of our nation. Here, however, we may again
select a few outstanding names. Becket, in many ways
the most famous of our archbishops, was born in London,
behind the Mercers' Chapel in the Poultry, the son of
a wealthy merchant. Colet, Dean of St Paul's, was the
founder of one of the most noted City Schools. John
Wesley was educated at the Charterhouse, and held his
first Methodist Conference in London in 1744. Sydney
Smith, the witty Canon of St Paul's, had previously
been preacher at the Foundling Hospital, and at Berkeley
Chapel.

The statesmen who have identified their fortunes
with London are very numerous. Thomas Cromwell
was born at Putney, and, after helping Henry VIII in
the dissolution of the monasteries, was beheaded on
Tower Hill. Edmund Burke received his legal training
at the Middle Temple, and his speech in Westminster
Hall on the impeachment of Warren Hastings was one
of his greatest efforts. He was fond of London, and
among his literary friends were Dr Johnson and Goldsmith.
William Pitt, one of our foremost prime ministers, has a
statue to his memory in the Guildhall. Sir Robert Peel
has special claims on our notice, for not only did he repeal
the Corn Laws, but he formed the Metropolitan Police
Force, or "Peelers," as they were once termed, whose
members took the place of the old watchmen. He was

thrown from his horse near Hyde Park Corner, and died from the effects of the fall at his house in Whitehall. Peel's services to London are brought to our mind by his statue at the west end of Cheapside.

Charles George Gordon

Among the men of action whose fortunes were connected with London, we will name the two foremost in our history. Nelson, the greatest of our admirals, and Wellington, the hero of a hundred fights, are assuredly the pride and possession of the Empire, and

London honoured them both with public funerals, which were unique in their pomp. Both of them were laid to rest in St Paul's, where there are splendid monuments to their memory. Nelson's monument in Trafalgar Square is one of the sights of London; and everyone knows the equestrian statue of the Duke of Wellington in front of the Royal Exchange. The Duke lived at Apsley House, Piccadilly, and when he was out of favour with the London mob, the windows of that residence were broken. General Gordon was born at Woolwich, and he has a statue in Trafalgar Square. Here we may mention that London has been very generous in raising statues to its military heroes, and Havelock and Napier, as well as the Crimean soldiers, and those who fought in South Africa and elsewhere are all commemorated either in the public squares, or in the cathedrals and churches of the metropolis.

The historians and antiquaries who flourished in London form a goodly company. Leland, who was born about 1506, in London, was educated at St Paul's School. Among his works *The Itinerary* is the best known, and shows that in the early Tudor days men were beginning to take an interest in the past history of their country. John Stow, who formed a worthy successor to Leland, was the son of a tailor of Cornhill, where he was born in 1525. Stow's *Survey of London and Westminster* is the foundation of all later work on that subject. A very striking monument to his memory is in the church of St Andrew Undershaft. William Camden, scholar, antiquary, and historian, was educated at Christ's Hospital,

Stow's Monument, St Andrew Undershaft

and St Paul's School. He stands out as the great historian of his country in the reign of Elizabeth, and his *Britannia* does for the whole country what Stow did for London. John Strype, who lived in the later Stuart period, received

Chaucer

his education at St Paul's School. He is connected with Hackney, and continued Stow's *Survey of London and Westminster* to the beginning of the eighteenth century.

London has always been famous for its poets, and during the Elizabethan period it was "a nest of singing-

birds." One of our earliest poets, John Gower—
"Moral" Gower as he was called—lies buried in South-
wark Cathedral, where there is an effigy to his memory.
Chaucer, too, has made Southwark famous for all time.
He was born in London, resided in Aldgate, and became
Comptroller of the Petty Customs. It was at the Tabard
Inn, Southwark, that his company of pilgrims assembled
for their journey to Canterbury. Chaucer subsequently
lived in Westminster, and was buried in the Abbey.
Edmund Spenser, author of the *Faerie Queene*, was born
in London, probably near the Tower. He ended his
life in distressed circumstances, and died in King Street,
Westminster. Shakespeare, the greatest of all our
names, came to London when he reached manhood, and
his first work there was probably some menial office in
connection with the Curtain Theatre at Shoreditch. All
his work in London was either as actor or playwright,
and the Globe and the Blackfriars Theatres on the
Surrey side were the scenes of his triumphs. His con-
temporaries in London were Beaumont and Fletcher,
Massinger and Ben Jonson, and the latter he often met
at the Mermaid Tavern in Bread Street. His friend,
Ben Jonson, realised his greatness, for he says that Shakespeare wrote "not for an age but for all time." There
is a fine monument to Shakespeare in the Abbey, and
a later one in Leicester Square. John Milton, the greatest
of the Stuart poets, the son of a London scrivener, was
born in Bread Street, Cheapside. He was educated at
St Paul's School, and buried in St Giles's, Cripplegate,
where there is a monument to him in the churchyard.

Among the London poets of the early Georgian period
Dryden, Pope, and Gray take a high place, but Goldsmith
has stronger claims on our attention. He reached London
in destitution, became a physician in Southwark, and then
usher in a school at Peckham. Eventually he found the
friendship of Dr Johnson, was a member of the famous

John Milton

Literary Club, died at 2, Brick Court, Temple, and was
buried in the Temple Church. Of the Victorian poets,
Robert Browning was born at Camberwell, educated at
Peckham and at University College, and buried in Westminster Abbey. Thomas Hood, who sang *The Bridge of
Sighs* and the *Song of the Shirt*, was born in the Poultry
quite close to Bow Church.

The men of letters who have made London their home are even more numerous than the poets. Pepys, writer of the famous *Diary*, was educated at St Paul's School, and lived in Buckingham Street, Strand. He was

Samuel Pepys

buried in St Olave's, Hart Street, at nine o'clock at night and there are monuments in that church to him and to his wife. For a correct and realistic knowledge of London of the time of Charles II we must turn to the pages of this diarist. Evelyn, too, who wrote a diary of the same period, lived at Deptford, and his work is often useful to check Pepys' statements. Nor must we forget Daniel Defoe, who wrote the *Journal of the Plague Year*. He was born in St Giles's, Cripplegate, in 1661, and

Daniel Defoe

is buried in Bunhill Fields Cemetery. Of all the men of letters of whom London is justly proud, none is greater than Dr Johnson. He is styled the "leader of literature in the eighteenth century," and the best part of his life's work was accomplished in London.

The celebrated Club founded by Sir Joshua Reynolds
at the Turk's Head, Gerard Street, Soho, included
among its members Johnson, Burke, Goldsmith, Gibbon,
and Boswell, but Johnson was the acknowledged leader.

Dr Johnson

London was the greatest place in the world to Johnson.
"Fleet Street," he once said, "has a very animated
appearance ; but I think the full tide of human existence
is at Charing Cross." The house in Gough Square,
where he compiled his *Dictionary*, still stands ; his seat in

Dr Johnson's seat at the " Cheshire Cheese "

St Clement Danes Church in the Strand has a brass
plate on it ; and the " Cheshire Cheese," one of his

Dr Johnson's House, 17, Gough Square

favourite haunts in Fleet Street, is yearly visited by
thousands of his admirers. Johnson died at 8, Bolt

Court, Fleet Street, and was buried in Westminster Abbey. Happily, we can follow him in his London life, for Boswell's biography gives us the minutest details of the man and his friends. Charles Lamb, the gentle

Charles Lamb

essayist, was a true Londoner. Born in the Middle Temple, he was educated at Christ's Hospital, and was a clerk in the East India House for thirty-six years. Most of his essays were written in London before he retired to Enfield.

When we come to the Victorian era, we have a goodly company of men of letters who delighted in London.   Thackeray, the great novelist, was educated

William Hogarth

at the Charterhouse, and *Vanity Fair*, *Esmond*, and *Pendennis* were written at Young Street, Kensington, or Kensington Palace Road where he died.   Dickens,

T. Rowlandson.
The Caricaturist.

Thomas Rowlandson

too, was essentially a Londoner. He knew the metropolis of his time as few others knew it, and we find in such novels as *David Copperfield*, *Sketches by Boz*, and the *Pickwick Papers* evidences of this intimate

George Cruikshank

knowledge. He lived at 48, Doughty Street, and in the Marylebone Road. James Mill, and his son John Stuart Mill lived in London, the latter being born in Rodney Street, Pentonville.

London has been the home of some of our greatest

painters. Hogarth, born in Bartholomew Close, Smith-field, and apprenticed to an engraver in Cranbourne Street, knew all the phases of London life, and reflected them in his pictures. Thomas Rowlandson, the great master of caricature, was born in the Old Jewry, and Sir David Wilkie, whose paintings of humble life are so familiar, lived in about a dozen houses in various parts of London. George Cruikshank's humorous pencil illustrated Dickens's books and was ever ready to assist in the war against intemperance.

From this brief survey of London's Roll of Honour, it will be seen that the City has always had a charm, almost a fascination, over the lives of many of our great men. An attachment for London is the experience of most people who come to it early enough, and Dr Johnson expressed this feeling when he said : " Why, sir, you find no man at all intellectual who is willing to leave London. No, sir, when a man is tired of London he is tired of life, for there is in London all that life can afford."

28. THE CITY OF LONDON AND THE BOROUGHS IN THE NORTH-EAST AND SOUTH-EAST OF THE COUNTY OF LONDON.

The City of London is one of the smallest divisions of the County of London, and has the smallest population. Neither its size nor its night population, however, gives an indication of its great importance. The City is the very heart of London, and the centre of the commercial life of the Empire. It is also a County and has enjoyed exceptional privileges from the earliest times. Although it has a night population of only 19,657, the resident daily population is over 300,000, while it is calculated that upwards of a million people enter it and leave it in 24 hours. Before 1907 it was divided into 112 parishes, but in that year they were consolidated into one parish. The City has 25 Wards, and is governed by the City Corporation, which consists of the Lord Mayor, 25 Aldermen, and 206 Common Councillors. Among the most important public buildings in the City are the following. The Guildhall is the Council-hall of the City, and the Mansion House is the residence of the Lord Mayor. The Central Criminal Court is a fine modern building which has taken the place of the Old Bailey. The Bank of England is the centre of the banking business of the country. The Royal Exchange and the Stock Exchange are devoted to the commercial and monetary life of London. The General Post Office has all its chief work concentrated in various large buildings known as St Martin's-le-Grand. St Paul's Cathedral is the most important church in the City, and is of national interest as the burial-place of great soldiers, sailors, and other men of note.

Bermondsey consists of the civil parishes of Bermondsey, Rotherhithe, Horsleydown, St Olave, and St Thomas, Southwark. In point of size this borough is the nineteenth in the County of London, and its density of population shows about 84 persons per acre. The growing population of recent years has had fewer houses to live in, owing to the increase of warehouses and factories. Within the borough there are several open spaces with a total acreage of 75. The chief of these spaces is South-wark Park. The borough is divided into 12 wards, and the borough council has 9 aldermen and 54 councillors. Bermondsey has long been noted for its tanneries, where leather of the best class is manufactured. In 1082 a monastery was founded at Bermondsey for monks of the Cluniac Order. The site of the monastery was granted at the Reformation to Sir Robert Southwell, and soon afterwards a fine mansion was built on it. No traces now remain of the monastery, although in the early nineteenth century its ancient gate, with a large arch and postern on one side, was standing. Bermondsey has a handsome Public Library, and a good Museum in which there is a fine collection of flint implements, and also a large number of exhibits illustrating the tanning and leather industry.

Bethnal Green is one of the smallest of the London boroughs, but has a dense population, there being as many as 170 persons to the acre. A very large proportion of the people live in tenements. The borough has about 100 acres of open spaces, including a portion of Victoria Park. The borough council consists of 5 aldermen and 30 councillors, and the borough is divided into 4 wards. Bethnal Green was formerly chiefly inhabited by weavers, and even now there are silk manufactures. On June 26, 1663, Pepys has the following entry in his *Diary*:—"By coach to Bednall Green to Sir W. Rider's to dinner. A fine merry walk with the ladies alone after dinner in the garden; the greatest quantity of strawberries I ever saw, and good. This very house was built by the Blind Beggar of Bednall Green, so much talked of and sung in ballads."

Camberwell is the fourth among the London boroughs in area, and fifth in population. The extent of the borough is the same as that of the civil parish, and it is divided into 20 wards. The population is unevenly distributed, and is most dense in the North division. There are 58 persons to the acre in Camberwell. The borough is fortunate in having two large open spaces within its area, viz. Peckham Rye and Dulwich Park, besides 14 smaller ones. The borough council consists of 10 aldermen and 60 councillors, each of the wards being represented by 3 councillors.

Dulwich is famous for its College called "God's Gift College in Dulwich." This was built and endowed in 1619 by Edward Alleyn, a celebrated actor and Master of the King's Bears. He founded it as a chapel, a school house, and 12 almshouses, and so valuable has the property become which he left for its endowment, that a new and conspicuous building was erected in 1870 at a cost of £100,000. The Old College is a quaint building, and has attached to it a gallery with a fine collection of pictures.

Deptford is one of the smaller London boroughs, and its population shows a density of 70 to the acre. The borough has only a small area of open spaces, of which the chief are Deptford Park and Telegraph Hill. Deptford has 6 aldermen and 36 councillors on the borough council, and is divided into 6 wards. There are engineering works here, and some manufactures of earthenware, soap, and chemicals. The old Naval Dockyard was filled up in 1869, and converted into a foreign cattle-market, which belongs to the City Corporation. The Royal Victualling yard is still maintained, and here cattle are slaughtered, and biscuits and other provisions are stored for the Royal Navy. Sayes Court belonged to Evelyn, and was the residence of Peter the Great while he was studying shipbuilding at Deptford. Captain Cook's ships, the *Resolution* and the *Discovery*, sailed from Deptford, and it was here that Queen Elizabeth knighted Sir Francis Drake.

Finsbury is the smallest borough in London save Holborn. It consists of the civil parishes of Charterhouse, Clerkenwell, Glasshouse Yard, St Luke, and St Sepulchre. In general character the various parts of the borough are somewhat similar. They are all fully built upon and densely populated, the number of persons per acre being about 150, as compared with about 60 for all London. The population is now declining, owing to the decrease in the area available for residence, which is becoming more and more limited by the increase of business premises. As regards open spaces, Finsbury is very deficient, there being only about 16 acres of breathing space for the large population. The borough is divided into 11 wards, and the borough council consists of 9 aldermen and 54 councillors.

Greenwich is the sixth London borough in area, and its population shows a density of 25 persons to the acre. It contains within its boundaries several important open spaces, such as Greenwich Park, and parts of Blackheath and Woolwich Common, which cover an area of 390 acres. The borough consists of the parishes of Charlton and Kidbrooke, St Nicholas, Deptford, and Greenwich, and it is divided into 8 wards. The borough council has 5 aldermen and 30 councillors. Greenwich possesses several important manufactories, including telegraph works, engineering works, and chemical works. The Royal Observatory in Greenwich Park has a world-wide celebrity: geographers calculate longitude from its meridian, and chronometers are sent here to be tested.

Greenwich Hospital, now the Royal Naval College, is a splendid edifice, built upon the site of a royal palace, which was the birthplace of Henry VIII, Queen Mary, and Queen Elizabeth. The Painted Hall contains some good naval pictures and statues.

Blackheath lies to the south of Greenwich Park and is crossed by the ancient Watling Street leading to Dover. Here were encamped the rebels under Wat Tyler (1381) and Jack Cade (1450), and here the citizens of London welcomed Henry V on his return from Agincourt.

Hackney is the seventh among the London boroughs in point of size and sixth of population. The borough coincides with the civil parish, and is divided into 8 wards. There is an area of 618 acres of open spaces, which include Hackney Marsh, Hackney Common, Hackney Downs, and London Fields. The borough council consists of 10 aldermen and 60 councillors. Hackney was formerly the residence of the Vere, Rich, Zouche, Brooke, and Rowe families, and is often alluded to by the old dramatists. It is no longer a fashionable district, but there are some good old-fashioned houses, with fine trees and gardens. How Hackney has changed may be gathered from one entry in Pepys' *Diary* of May 12, 1687. It is as follows:—"Walked over the fields to Kingsland and back again; a walk I think I have not taken these twenty years; but puts me in mind of my boy's time, when I boarded at Kingsland, and used to shoot with my bow and arrows in these fields."

Islington has the largest population of all the London boroughs, and is eighth in point of size. The borough coincides with the civil parish, and is divided into 11 wards. Islington is densely populated, having no fewer than 105 persons to the acre. It is not well off as regards open spaces, which, including Highbury fields, have an area of only 40 acres. The borough council consists of 10 aldermen and 60 councillors

Brooke House, Clapton, Garden front

(*The residence of the Brooke family*)

The Agricultural Hall, the second largest building of the kind in London, is in Upper Street, and here every December is held the Cattle Show. The Angel Hotel, an inn famous since the seventeenth century, was depicted in Hogarth's picture "The Stage Coach," but has since been much altered and modernised. Islington was formerly famous for its ducking-ponds, its cheese-cakes, and custards, and was often mentioned in works of the sixteenth and later centuries. Dr Johnson, referring to the dairy farms at Islington, says: "A man who gives the natural history of the cow is not to tell how many cows are milked at Islington"; and William Collins the poet, Oliver Goldsmith, and Charles Lamb, among many other eminent names, were connected with Islington.

Lewisham, on the Ravensbourne, is the third London borough in point of size, and is seventeen times as large as Holborn, which is the smallest borough. At present this large area is only thinly populated, there being about 22 persons to the acre. The borough has 266 acres of open spaces, which comprise a portion of Blackheath and the Hilly Fields. The borough is divided into 10 wards, and the council has 7 aldermen and 42 councillors. Included in the borough are Lee, Blackheath, Brockley, Catford, Forest Hill, and Sydenham. The Horniman Museum is at Forest Hill, and the Crystal Palace is near Sydenham. The parish church of Lewisham is interesting as an example of the Classical style, and Colfe's Grammar School was founded and endowed in 1656.

Poplar consists of the parishes of Bow, Bromley, and Poplar, and is the eleventh London borough in point of size. Its population is large, there being about 70 persons to the acre. The borough has several small open spaces, and 72½ acres of Victoria Park. The borough council consists of 7 aldermen and 42 councillors, who represent the 14 wards into which the borough is divided. Poplar is on the River Thames between Blackwall

Reach and Limehouse Reach, and contains the East and West India Docks and Millwall Docks.

Shoreditch, one of the smallest of the London boroughs, has a dense population of 170 persons to the acre. The population is decreasing, owing to several causes, but chiefly to the fact that dwelling-houses are being pulled down to make way for factories, etc. The few open spaces for this congested neighbourhood only amount to 6½ acres. There are 8 borough wards, and the council consists of 7 aldermen and 42 councillors. Shoreditch is a typical East End district. Its name is said to be derived from a popular ballad about Jane Shore repenting of her sins in a ditch! In the sixteenth century its theatres "The Curtain" and "The Theatre" were for some time the only two in London. It is said that Shakespeare stood at the doors of the Shoreditch playhouses and held the horses of the spectators during the performance. His *Romeo and Juliet* was performed in 1597 at "The Curtain."

Southwark ranks as one of the smaller boroughs, but it has one of the largest populations. It has no less than 170 persons to the acre, and is thus the most densely populated area in London. It consists of the parishes of Christchurch, Southwark, Newington, St George the Martyr, Southwark, and St Saviour, Southwark. There are only 11 acres of open spaces in the borough, and Kennington Park is the nearest of the large open spaces. The borough has 10 wards, and the council consists of 10 aldermen and 60 councillors. The historical interest of Southwark is considerable. Its early importance arose from the fact that it was on the line of Watling Street, and near the great crossing-place on the southern bank of the Thames. Its priory was closed at the Reformation, but the priory church now forms he Cathedral Church of the See. The Tabard Inn was connected with Chaucer's *Canterbury Tales*, and some of the theatres were associated with Shakespeare and other Elizabethan actors and

Gower's Tomb, Southwark Cathedral

dramatists. Guy's Hospital is in this borough, and there is a large market. Southwark has many industries and manufactures, e.g. glass-making, mat-making, brewing, and tanning.

Stepney, the fourth London borough in point of population, has a density of 160 persons to the acre. This borough includes the following parishes:—Limehouse, Mile End, Norton Folgate, Old Artillery Ground, Ratcliff, St Botolph, Aldgate, St George in the East, Shadwell, Spitalfields, Wapping, and Whitechapel. The Tower of London is within the area of the borough. Stepney has very few open spaces, and their area is only 48 acres. The churchyard of St Dunstan, seven acres in extent, forms a veritable oasis among the mean brick houses. The population of Stepney is the most cosmopolitan in London. At the 1901 census there were upwards of 54,000 foreigners in this borough, or about one-fifth of the whole population. Every country in Europe was represented, and the aliens from Russia, Poland, Germany, and Austria were very numerous. There are 19 wards in the borough, and the council consists of 10 aldermen and 60 councillors. The wide Whitechapel and Mile End Road is the main thoroughfare in the north, while the Commercial Road and East India Docks are in the south. The finest church in Stepney is St Dunstan's, a handsome Perpendicular building. Colet, the well-known Dean of St Paul's, was once the vicar. In Spitalfields are still some descendants of the old Huguenot refugees who settled here in 1685 and introduced silk-weaving into London.

Stoke Newington is among the smallest of the London boroughs and, except the City of London, has the least population. South Hornsey, which forms a part of the borough, is fully built upon, but there is room for a slight increase in Stoke Newington. This borough has the beautiful Clissold Park of 57 acres, and in the neighbourhood is Finsbury Park of 115 acres. There are 6 wards in Stoke Newington, and the council has 5 aldermen and 30 councillors.

Woolwich is the second largest of the London boroughs, and the least densely populated, for there are only 14 persons to the acre. It comprises three parishes, Woolwich, Plumstead, and Eltham, and has no less than 355 acres of open spaces, which include Plumstead Common, Bostall Heath and Woods, Eltham Common and Green, and Woolwich Common. There are 11 wards in the borough, and the council consists of 6 aldermen and 36 councillors. There is a free ferry from Woolwich pier to North Woolwich, which is a detached portion on the Essex side of the river. The chief feature of Woolwich is the Arsenal, one of the most extensive and complete in the world. Among other buildings of interest are the Military Academy and the Herbert Hospital.

A Table giving the Area and Population of the City
of London, and the Boroughs in the North-East and
South-East of the County of London.

THE CITY OF LONDON, AND BOROUGHS	AREA IN ACRES	POPULATION IN 1911
CITY OF LONDON	672·7	19,657
Bermondsey	1499·6	125,960
Bethnal Green	759·3	128,282
Camberwell	4480·0	261,357
Deptford	1562·7	109,498
Finsbury	589·1	87,976
Greenwich	3851·7	95,977
Hackney	3288·9	222,587
Islington	3091·5	327,423
Lewisham	7014·4	160,843
Poplar	2327·7	162,449
Shoreditch	657·6	111,463
Southwark	1131·5	191,951
Stepney	1765·6	280,024
Stoke Newington	863·5	50,683
Woolwich	8276·6	121,403
	41832·4	2,457,533

Note. The Administrative County of London including the
City of London had a total area of 74,839 acres and a population
of 4,522,961 at the census of 1911.

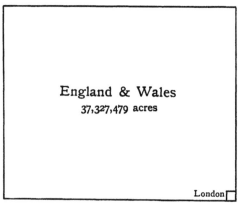

Fig. 1. Area of the Administrative County of London (74,839 acres) compared with the area of England and Wales

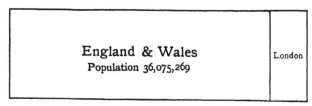

Fig. 2. The Population of the Administrative County of London (4,522,961) compared with that of England and Wales in 1911

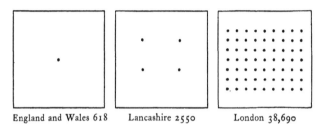

| England and Wales 618 | Lancashire 2550 | London 38,690 |

Fig. 3. Comparative Density of Population to sq. mile
in 1911

(*Each dot represents 618 persons*)

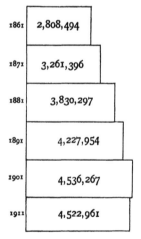

1861	2,808,494
1871	3,261,396
1881	3,830,297
1891	4,227,954
1901	4,536,267
1911	4,522,961

Fig. 4. The Growth of Population in London from
1861—1911

INDEX

Milton Keynes UK
Ingram Content Group UK Ltd.
UKHW032321161024
449665UK00001B/5

9 781107 667501